A complete guide to collecting

Antique Pipes

Benjamin Rapaport

Revised

4880 Lower Valley Rd. Atglen, PA 19310 USA

Dedicated to
all silent pipe collectors . . .
that the joys of pipe collecting
not go unsung!

Price Guide: 1998
Copyright © 1979 Schiffer Publishing Limited.
Library of Congress Catalog Card Number: 79-67024

All rights reserved. No part of this work may be reproduced or used in any forms or by any means—graphic, electronic or mechanical, including photocopying or information storage and retrieval systems—without permission from the copyright holder.

ISBN: 0-7643-0596-4
Printed in United States of America

Published by Schiffer Publishing Ltd.
4880 Lower Valley Road
Atglen, PA 19310
Phone: (610) 593-1777; Fax: (610) 593-2002
E-mail: Schifferbk@aol.com
Please write for a free catalog.
This book may be purchased from the publisher.
Please include $3.95 for shipping.

In Europe, Schiffer books are distributed by
Bushwood Books
84 Bushwood Road
Kew Gardens
Surrey TW9 3BQ England
Phone: 44 (0)181 948-8119; Fax: 44 (0)181 948-3232
E-mail: Bushwd@aol.com

Please try your bookstore first.
We are interested in hearing from authors
with book ideas on related subjects.

Preface

There is no single nation on this earth not yet familiar with tobacco. About 150 years ago, in 1834 to be exact, a book entitled HINTS ON ETIQUETTE, stated: "If you are so unfortunate as to have contracted the low habit of smoking, be careful to practice it under certain restrictions; at least, so long as you are desirous of being considered fit for civilized society." Today, this habit is quite part of our civilized world. Collecting smoking paraphernalia is indicative of a civilized world reflecting on its own cultural past.

Tobacco has been cited as a therapeutic in medical tracts, as a source of wealth in commercial and agricultural accounts of the history of nations, as a cultural phenomenon in books on social etiquette, as a rite in cult studies, as a consumption model in marketing and consumer surveys.

Tobacco consumption has taken many forms—as the cigar, the cigarette, snuff, plug and in the pipe. The pipe, a phenomenon of the Seventeenth Century, diminished in popularity in the Eighteenth Century with the introduction of snuff, cigars and cigarettes and regained its stature after having adjusted its appearance and its substances to meet changing demands.

In today's explosive world of literature, illustrated books and guides on tobacco accessories and paraphernalia have not been exempt. In the last decade, some fifty new treatises in English and foreign languages have brought a renaissance in the study of smoking and its collectable utensils—cigar store Indians, cigar moulds, cigar boxes and box labels, cigar bands and cutters; cigarette packages, cigarette and other trade cards; tobacco tin tags, match boxes, matchbook covers and match safes; snuff rasps, mulls, boxes, jars and bottles; tobacco jars, boxes and tampers. **ANTIQUE PIPES** fills the void of absent information about another increasingly popular collectable, the antique smoking pipe. I say absent information for there is no single national repository of archives, chronicles or annals of the pipe industry. There is only one American periodical, *The Pipe Smoker's Ephemeris,* which acts as an occasional clearing house, an information medium for pipe collectors. To complicate matters, pipe collectors, as a rule, enjoy their privacy and anonymity; this further inhibits the exchange of information. Twenty years of reading, research and world travel have furthered my own antique pipe knowledge. Having compiled **A TOBACCO SOURCE BOOK** in 1972 and several monographs on tobacciana between then and now, I am confident that **ANTIQUE PIPES** will satisfy anyone's curiosity, because it provides, in a single textual and illustrated guide, definitive information heretofore not readily available to the average collector about carvers, the provenance of pipes and early guilds and factories of Europe and America.

Encouragement was not wanting. I owe the initial credit to Thomas Allan Dunn, Founder and President of the Universal Coterie of Pipe Smokers, Publisher and Editor of that infrequent quarterly, *The Ephemeris,* and my mentor; he planted a seed some ten years ago when I was introduced to his circle of Coterians. John B. Stone, a fellow pipe collector and friend, was supportive of the idea throughout its germination. The catalyst was Peter B. Schiffer, President of Schiffer Publishing Ltd., who called one evening in December, 1978, and offered me the opportunity to author this book. He travelled with me to photograph public and private collections and lived through the tedium of manuscript editing.

Persistence, a good set of foreign language dictionaries and myriad correspondence with loyal, collaborative and enthusiastic collectors, however, were the key ingredients. To those collectors and to the many institutions who elected to participate in this endeavor and exhibit their prized possessions in this book, I express a hearty vote of thanks; their names appear in appropriate places throughout the book. There is also my wife, Liz, who has never really understood my pipe-o-mania, but has encouraged and stimulated me through pipe-smoke thick and budget-thin times so that I might pursue this hobby and enlarge my collection. She typed the many revised drafts with diligence and patience. Without her commitment, the target date for publication would never have been met.

ANTIQUE PIPES is dedicated to all who see the aesthetic quality found in antique pipes. It is written for the appraiser, the antique dealer, the student of tobacco culture, the historian and especially for the established and would-be pipe collector.

Contents

CHAPTER		PAGE
	PREFACE	iii
1.	INTRODUCTION	7
2.	THE CLAYS	15
3.	THE PORCELAINS	35
4.	THE MEERSCHAUMS	49
5.	THE WOODS	93
6.	ETHNOGRAPHICA I: NEAR AND FAR EAST	123
7.	ETHNOGRAPHICA II: AMERICA AND AFRICA	151
8.	METAL MISCELLANEA AND COGS	173
9.	YESTERDAY'S INVETERATE COLLECTOR—ART, INVESTMENT OR BOTH?	191
10.	THE PUBLIC AND PRIVATE PIPE	201
11.	RESTORATION AND REPAIR	227
12.	RECOMMENDED READING	237
	SELECTED BIBLIOGRAPHY	243
	INDEX	249

Values: Prices indicated reflect and take into consideration the current market values in both the United States and Europe.

Kaiser Wilhelm II was the protector of the Wilhelm Imhoff pipe company, Cassel, Germany and appeared as here on the back cover of the company's catalog, c. 1912.

Chapter 1

INTRODUCTION

Today, across the nation and overseas, there is a revival in the tobacco industry to craft pipes not just as a utensil for smoking, but also as an extension of man's creative ability. There is a new breed of carver in Turkey sculpting fascinating and intricate meerschaums for export which approximate the quality of the once-esteemed meerschaums of Vienna, Hamburg, Prague and Budapest. There is a renaissance in briar conceived by a handful of adept artisans in the United States and Canada, shaping ebauchons, briar burls, with personality and individuality; a few create only to special order, others design theirs for general consumption through established retail outlets. In England and on the Continent, old moulds have reappeared and with them a revival in clay pipe reproductions: the cutty, the churchwarden, the figural, the personage pipe.

Specialty shops, which recognize the need to cater exclusively to masculine hobbies have been successfully marketing tobacco collectables; of these, a few come to mind: Mantiques and Ampersand in New York City; The Old Order, Thiensville, Wisconsin; The Pipesman, Milwaukee, Wisconsin; Brian Tipping's Pipe Shop, King's Road and Astley's, Jermyn Street, London; Denise Corbier's Pipes Anciennes, Rue de l'Odeon, Paris; Collector Clay Pipes Co. Ltd., Vancouver, B. C., Canada.

In just the past few years, I have witnessed a keener interest by public museums to acquire tobacco artifacts for prime location permanent display. The United States tobacco industry has begun to emulate the European attitude toward and enthusiasm for the cultural and historical significance of tobacco utensils. I now read of large corporate expenditures to search out memorabilia, ephemera, smokers' paraphernalia and other relics of past-time smoking customs for public exhibit.

Tobacco bibliophiles have witnessed a recent surge in the demand for the authoritative works on tobacco culture. Many Seventeenth to early Twentieth Century books of prose, poetry and anecdotes of smoking have reappeared as facsimiles or reprints to supplant original editions, now out-of-print.

Lastly, today's affluence and increased leisure time have nurtured a geometric rise in antique and collectable enthusiasts. Among them are a number of tobacciana aficionados, especially antique pipe collectors.

What does all this signify? It evinces, to me, the acceptability of smoking relics as part of our culture and heritage. It indicates that art lovers have come to recognize and relish the beauty and the elegance of tobacco implements, specifically the pipe. Pipe collecting, however, is a hobby only for the investigative and inquisitive.

The common thread between all established pipe collectors is the constant cry for information. There is too little precise knowledge, too little documentation upon which to establish exact and deducible provenance on pipes. The lineage of most antique pipes is shrouded in myth, mystery and misinformation. The clay is perhaps the only exception.

As one example of the few records of the industry, Ruhla, in Thueringia, Germany, was a principal pipe manufacturing center in the Nineteenth Century. The following represent annual pipe production statistics for the period 1850–1870:

Meerschaums	570,000
Imitation Meerschaums	500,000
Porcelain bowls	9,600,000
Woods	5,000,000
Clays	3,000,000[1]

Nineteen million pipes were produced in one small German town alone in a short period of twenty years.

There are, fortunately, a few accepted generalities. First, there were specific European and American centers for pipe manufacturing. A smoking or storage case is proof of a pipe's origin; without one, material and style may be sufficient to suggest origin and confidence is peaked if the pipe is autographed, although that was rare. The hallmarks found on gold and silver pipe lids and collars contribute to dating pipes. A book on hallmarks, therefore, is an essential reference to a serious collector; it can provide an audit trail of the silver or goldsmith who made the mountings. The most difficult aspect is the why or wherefore of a pipe and therein lies every collector's frustration. Pipes have represented allegory and mythology, expressed a carver's whim or the capricious desires of a client, commemorated a specific historic national or personal event (birth, wedding, anniversary, death) or, with a facsimile bust, recorded for time immemorial, kings, queens and other heads of state, heroes, military leaders, artists and other famous personages. That is, to me, the fascination, the fantasy, the fun of pipe collecting!

Every material known to man has been used as a pipe or for pipes: stone, bone, reed, iron, tin, copper, bronze, brass, silver, gold and other precious metals, sundry woods, meerschaum, gourd, horn, corncob, clay and porcelain and any other matter which can be carved, moulded or shaped and retains heat.

The tobacco pipe was introduced into Europe from America sometime in the Sixteenth Century and spread throughout Central Europe in the Seventeenth Century. Small bowled pipes of single unit construction appeared first and were sized in relation to the high price of tobacco. As tobacco prices decreased, pipe bowl size increased. With the advent of the clay, the angular bowl carried a heel spur at its base, used principally as a setter, then as a place to incuse a manufacturer's hallmark or logo. As the clay spread across Europe, two developments took place. First, coloring and eventually the introduction of porcelain and pottery. Second, redesign, making the stem separable from the bowl. Other materials as metal, wood and meerschaum were later introduced, radically changing the utilitarian clay into elaborately decorated pipes of newer design, structure and ornamentation. Thus, the once functional pipe was transformed into the objet d'art.

Value, or price, is an important aspect for the collector and appraiser. Each pipe should be presumed authentic and should be evaluated on condition, size,

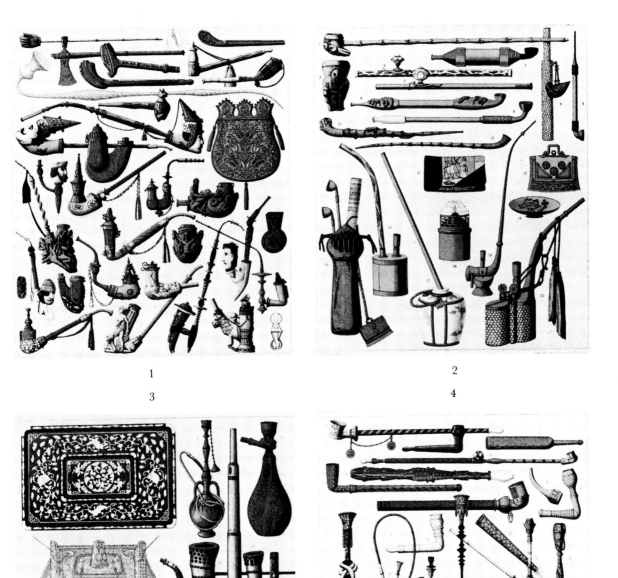

As an introduction to the collectible antique pipe, six polychrome lithographs from *Le Costume Historique* by M. A. Racinet, Libraire de Firmin-Didot et cie., Paris, 1888, are reproduced here. Racinet, an imaginative lithographer, sketched every conceivable artifact of the human race from the time of the Phoenicians forward. Pipes of Europe are found as Illustration 1. Illustration 2-4 are pipes of Asia, Illustration 5-6 are of African pipes. (Author's collection)

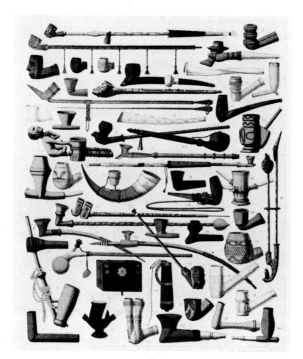
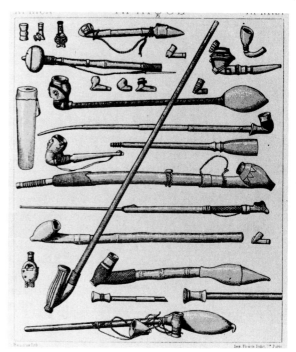

5 6

detail, ornamentation, execution, appeal and substance—a total and objective appraisal of the whole pipe. Where prices are shown in this book, they are only guides and are meant to give the reader an idea of what a person paid at the time purchased . . . the prices neither establish baselines nor delimit them. "Caveat emptor" is the prevailing caution. Valuation can only be done by a qualified appraiser.

A price guide could probably never be as thorough and final as the catalogued value of postage stamps, for example, because stamps, when issued, are practically all about alike, whereas pipes will vary greatly. Many were hand made, and as such they stand in a class by themselves. There have always been trends in pipes and pipe making and certain periods in history brought forth definite pipe styles. These should be studied and some sort of evaluation placed on pipes produced in any given era. Although there are certain types of furniture and other types of house furnishings traceable to a certain period of time, each separate piece will vary according to the quality of workmanship, present condition, and similar factors. Pipes should be treated in the same way. As more interest develops in pipes and pipe collecting, and as pipes of a century ago become more scarce, their prices will obviously go up.

One interpretation of pipe genres appeared in the form of cigarette cards. This set of 25 cards was issued by W. A. & A. C. Churchman (Imperial Tobacco Company) of England in 1926, entitled Pipes of the World. (Author's collection)

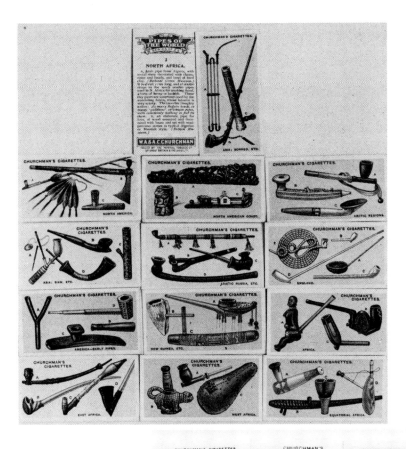
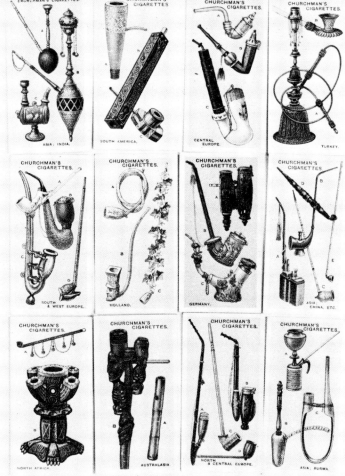

Though pipes take many forms as a national expression of many people, most antique pipe styles will appear in this photographic survey. The principal antique pipe classes are portrayed in chronological and historical perspective, highlighting interesting and colorful anecdotes. What might be a disappointment to many briar collectors is that the account terminates early in the Twentieth Century when standard shape briars and meerschaums became increasingly popular and the acceptable norm, when clays were essentially no longer in vogue and when the European porcelain pipe industry was in its nadir. It seemed to me opportune to curtail my discussion at just about World War I, being fully aware that today, avid briar collectors abound. The neglect is intentional. Without demeaning today's briar pipe, it is not, in my opinion, within the spectrum of antique smoking pipes. Chapter Nine recounts famous collectors and collections of the past and their contribution to having insured the collectable pipe elude obscurity. The book then proceeds with a list of locations where the pipe can be viewed publicly, some notes on pipe restoration and repair and lastly, books recommended for reading and further research.

Cigarette insert were also made in the United States. The Allen & Ginter Company of Richmond, Virginia issued these cards as a set of 50 in the last quarter of the 19th century. The series is called World's Smokers and 47 appear here. (Author's collection)

FOOTNOTES: CHAPTER 1

1. André Paul Bastien, La Pipe (Éd. Payot, Paris, 1973) n.p.

Chapter 2

THE CLAYS

"But sweeter still than this-than these-than all,
The pristine yard of Clay. It stands alone,
Fit for all climes, years, pockets large or small:
The height of smoking has been reach'd—all's known
By him who sticks to clay—no more he'll call
For pipe, save that the first to mortal shown.
A light for Adam's 'clay' was th'unforgiven
Fire which Prometheus filch'd from envious Heaven!
             From "Ode To My Yard of Clay"

L. W. L.

Scottish cutty, Irish Dudheen, Churchwarden, Alderman, Elfin, Fairy, London Straw, Yard—what are these? They represent members of the family in the long development of the humble, democratic, simple clay, the poor man's meerschaum, the historical pipe whose antiquity gained first place in the veneration of smokers! A 1745 dictionary amusingly defined the clay as a machine much used for smoking tobacco, consisting of a long slender tube or shank which is hollowed; made of baked clay with one end as a bowl or furnace for the tobacco, the fumes whereof are drawn out through the other end and so discharged; either long, short, plain, worked, white, varnished, unvarnished and of various colors. That definition is an almost verbatim account of the almost 200 years of clay pipe manufacture to that time.

It is accepted that clay pipes were manufactured in England as early as 1573 and by the turn of the century, they were in widespread use. Sadly, for that quarter century, records were generally not maintained and with certainty, most archaeologists agree that during the same period, clays were handmade. This whitish mineral, this mixture of aluminum silicate, or kaolinite and sand, reigned supreme until the end of the Nineteenth Century. But the link between tobacco and early clays was so universal that clay fragments were found in Mexico after Cortez's conquest in 1519, at early Indian mounds in America and throughout the Continent.

It is not my intention to retell the archaeologists' studies of the clay pipe industry or provide an anthropological audit for the myriad factories in England and on the Continent, although such information is pertinent and relevant. Clay collecting is the most fascinating, variegated and colorful pipe subset and therein will lie my emphasis in this chapter. It is fascinating for it is the most documented, the one about which collectors have the luxury of being or becoming most knowledgeable and expert—research has been so refined that, with confidence, a clay can be dated to within twenty years of its manufacture. It is the most variegated for the clay has been the most expressive in form, style and character. It is the most

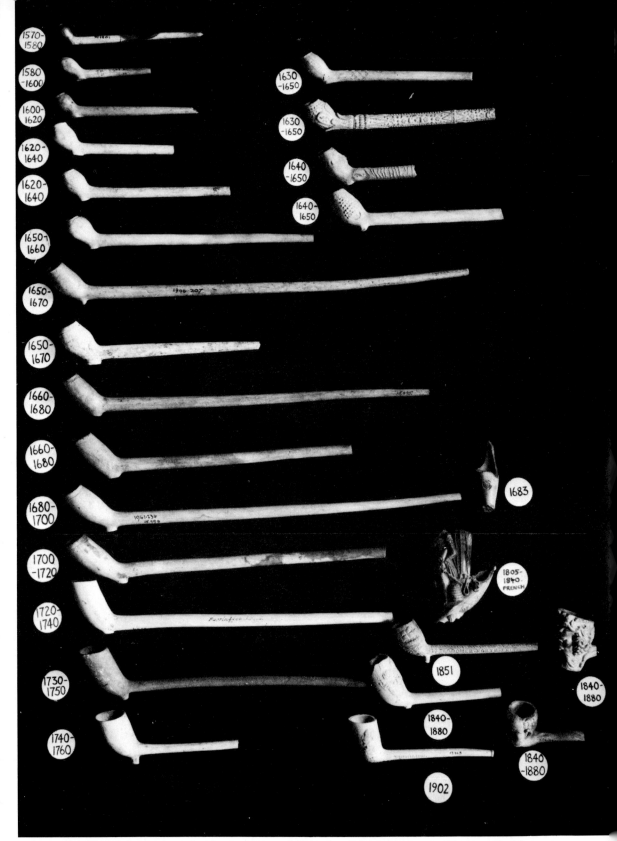

English clay development. A general overview of shapes, lengths and time frames. (Courtesy The Museum of London) $50-500.

colorful since the changes in clay bowl size, shank length and decoration were developments traceable to and an outgrowth of both the economics of smoking tobacco and advances in manufacturing processes and techniques.

The heyday of the clay pipe was essentially the century between 1650 and 1750, with the latter half of the Eighteenth Century giving rise to the popularization of snuff. Then, in turn, the porcelain and the meerschaum in the Nineteenth Century were too much competition for the frangible, low cost clay. In hot pursuit, clay styles attempted to emulate the meerschaum pipe and cheroot holder, but by 1900, when the briar found world-wide acceptability and cigar and cigarette smoking were custom, the clay was on its way out.

Adrian Oswald, a former keeper of the Department of Archaeology, Birmingham City Museum and a prolific writer on the history of clays in England, furnishes the following statistics of the principal English pipe-making centers:

Number of Makers[2]

	1600–1650	1650–1700	1700–1750	1750–1800	1800
Bristol	35	129	143	63	18
Chester	1	18	60	36	32
Hull	2	23	23	12	12
London	75	116	136	46	74
total:	113	186	362	157	136

According to Oswald, it is possible to trace as many as 3,400 clay makers in England alone![3]

As can be seen, after 1783 at least in England, both the loss of the American colonies as an export market and the Industrial Revolution had some effect on the marked decline in the clay industry.

From its inception, bowl size was directly related to the price and supply of tobacco; that is, when tobacco was prohibitively priced during the reign of Elizabeth I, clay bowls were exceptionally small, the pipe was unornamented and they sold for a few pence a dozen. These were called fairy pipes or cutties. As tobacco prices declined, bowls were made larger, stems longer and donned names as yard of clay, churchwarden and alderman.

Another feature which provides an archaeological audit is the pipe's base. Early clays were manufactured with a flat circular or flat oval heel. Gradually, the heel was transformed into a spur so customarily found on Nineteenth Century clays. Simple, straight-forward ornamentation on Seventeenth Century clays yielded to raised ornamentation in the Eighteenth Century; the Nineteenth Century was predominantly that of figural and enamelled clays.

It is on the Continent, however, where the most significant cosmetic changes to the clay occurred. "In 1620, the Dutch merchants were the largest wholesale tobacconists in Europe, and the people generally the greatest consumers of the weed."[4] This, no doubt, had some influence on dissenters in England who may have left for religious reasons or, for the purpose of this thesis, economic and pro-smoking reasons, since not only was tobacco taxed heavily but the English lived under the dreaded "Counter-blaste" reign of James I, an ardent anti-smoker. The first pipe-maker in Holland was one of those dissenters, William Baernelts or

A Dutch folianten—the suitor's or bridegroom's pipe. Folianten were made in lengths up to 100 centimeters and were often bedecked, as this one, with garlands of real or imitation flowers, ribbon, bows and moss. Its owner would present it to his sweetheart, begging for a light. If she complied the first time, it meant "maybe" to a marriage proposal. On the third offer, if she took this church-warden and lit it herself, that automatically implied a binding "yes" to a wedding date in the near future. After the wedding, the folianten was preserved in a special case and displayed on a wall in the home. This cased folianten bears the couple's initials and the wedding date 1902, with the bride's bouquet above the folianten below. (Courtesy Douwe Egbert's Pipe Room, Utrecht, Holland) $250-1000.

Barentz, who established his shop in Gouda in 1617. Other towns developed a clay pipe trade but by 1660, Gouda had become the center of Dutch clay manufacture with some 500 guild registered trademarks, using clay imported from Germany, Belgium and England. By the middle of the Eighteenth Century, the Gouda clay was at its pinnacle, offering short cutties, medium office pipes, about 22 centimeters long and long folianten, the bridegroom's pipe, a Dutch equivalent to the English churchwarden. During the same period, Belgium, Italy, Germany and Spain were clay pipe producing countries but none were as worthy as those from England and Holland . . . and eventually France.

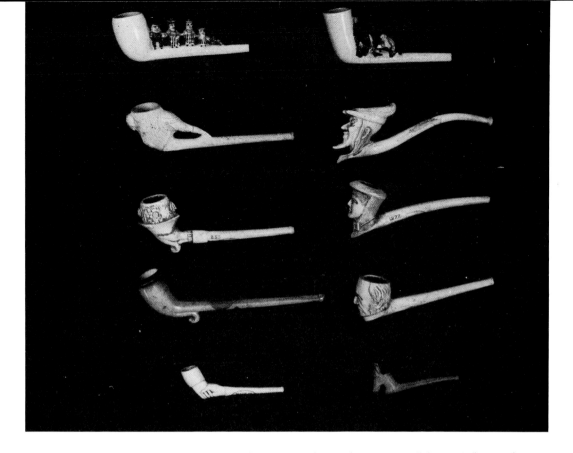

The infinite variety of the clay: plain, figural, ornament. Nineteenth century English or Dutch examples from various collectors. $25-75.

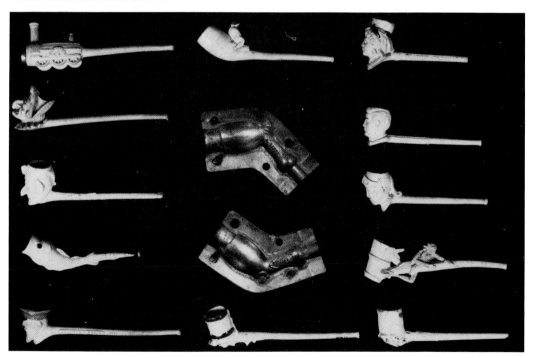

Nineteenth century English and French figurals from the US Tobacco Company Museum encircle a pipe mould. $25-100.

Two of the famous Gambier clay figurals: Disraeli on the left, General Boulanger on the right. (Courtesy le Musée du SEITA) Left: $50-125, right: $50-125.

A signed Gambier presentation bowl believed to be the image of the Roman queen of heaven Juno, the Greek goddess Hera. (Courtesy le Musée du SEITA) $125-300.

This unmarked 19th century presentation clay bowl of an emir or sultan is unquestionably French. (Courtesy le Musée du SEITA) $400-600.

By 1800 as in England, there was a decline in Dutch clay pipe factories—112 by official census. The decline was, in great measure, prompted by two significant events: the success of Germany's nonferrous, no chalk clay pipes and then France which, until the mid-Eighteenth Century imported clays from England and Holland but, with the rising popularity of pipe smoking, found sufficient demand to establish its own factories.

Initial French production is attributed to Avignon around 1670 and to one, Vausselin. Avignon city records indicate that in 1692, the brothers Van Slaton made clays with raised ornamentation. The premier French clay was produced on a very grand scale in the latter half of the Nineteenth Century at two major centers, St. Omer and Givet. The noted big three were: Duméril, founded in 1844-1845, Fiolet, founded in 1765, both of St. Omer and Gambier of Givet and Paris, established in 1780. There were other plants with lesser acclaim at 82 distinctive manufacturing locations in France. Minet and Roussel, as an example, was one early firm responsible for a kiln process to convert brown ferruginous clay to the more popular white, but the Fiolets and the Dumérils led the field by offering a legion of figural pipes, a major departure from the ornamental clays to that date. Today, clay collecting is limitless due to the decorative French—a panoply of ornamented single construction clays and the moulded heads of celebrities, mythological characters, animals and assorted others in the two piece clay with wood push stems.

In the Crystal Palace Exhibition catalog of 1851, both Fiolet and Duméril won honors: "Fiolet Louis of St. Omer (Pas de Calais)—Specimens of pipes made of clay. The exhibitor manufactures yearly above 200,000 gross of pipes either plain or varnished differing in size, form and length according to demand. They are made of 1,200 different shapes plain or ornamented, representing historical or fancy figures, animals, etc. By means of an enamel invented by the exhibitor brilliancy can be given to the plainest pipe."[5] And is it any wonder? "Two manufacturers alone sent out yearly from St. Omer forty-five millions of pipes made out of eleven thousand tons of clay."[6] These two firms were Duméril and Fiolet, with Fiolet the larger, employing some 700 workers of an estimated 5,000 employed in the French clay industry. In the opinion of one Oxford graduate in 1858: ". . . University men used to be rather particular about the pipes they smoked. The finest were made in France, and the favourite brand was 'Fiolet, Saint Omer.'"[7]

The name most synonymous with the ornamental clay pipe was Gambier of Givet and Paris. The firm truly capitalized on the technique developed in 1689 by Dr. Johannes Franz Vicarius, an Austrian physician who, while experimenting with glass-pipe stems, developed the first two-piece pipe, that is, separable head and stem.[8] As a result, Gambier pipe heads and other assorted clays, collated by Jean-Léo from the Gambier catalogues of 1894 and 1905 with occasional gaps in sequence, accounted for at least 2039 numbered varieties.[9]

But, as with the British and Dutch firms, Duméril shut down sometime around 1885, Fiolet in 1921 and Gambier in 1926. Professor Giusseppi Ramazzotti hints that there may have been a chance for the firm to survive, but the Germans had captured the moulds at Givet during the First World War and used the copper for guns.[10] I believe he meant to say brass, bronze or iron!

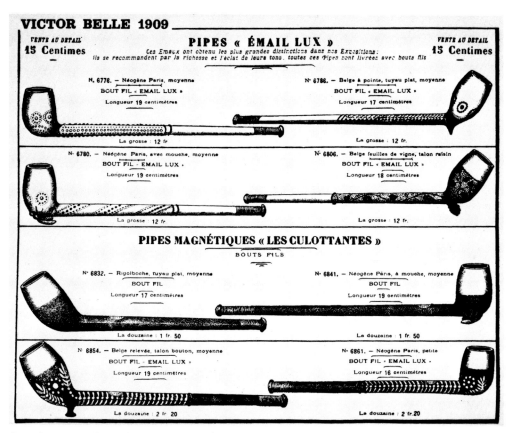

A catalogue page from Victor Belle, Erôme-Gervaud near Valence, France, in operation from 1834-1930. (Reproduction courtesy Jean-Léo)

The British remained close on the heels of the French and competed admirably well with quality figurals from firms as Charles Cropp and Sons and William Thomas Blake. One Blake advertisement of the day read: "Plain and Fancy clays, both long and short, Hunters, Crooks, Negro's, T.D.'s, Baltics, Lachlanders, Burn's Cuttys, Ben Nevis, Derry, Meerschaum washed London straws, etc., etc. certificate from the International Exhibition 1871 and medal 1873."[11] For those interested in the meerschaum washed clays, it was a British technique to copy the meerschaum. A kiln operator would coat the baked clay with a tasteless, innocuous preparation called meerschaum wash using a fine camel's hair brush, giving the pipe an appearance of real meerschaum.[12] Mr. Blake's export trade was to Australia, New Zealand, the West Indies, Canada, both coasts of Africa and South America. The clays destined for Africa's West Coast were tipped with bright red wax, "a colour it seems highly appreciated by our friends, the Fantees, Ashantees, and other coloured barbarians."[13]

For a moment, a digression to red clay. In France, the undisputed Nineteenth Century producer was Hippolyte Léon Bonnaud, Marseille. Founded in 1824, this firm specialized in marketing a colorful array of unglazed and glazed reddish brown figurals and straight-stemmers for more than a century until it closed its doors in 1955. In the East, two distinctive reds prevailed, both forerunners of today's coffeehouse clay. The earlier precursor was Debrecen, about 130 miles east

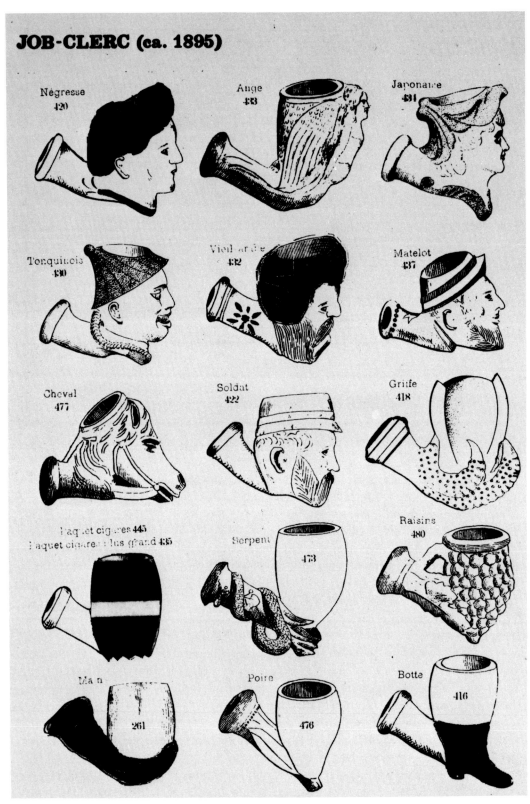

A two page extract of the Job Clere catalog of 1895. (Reproduction courtesy Jean-Léo)

JOB-CLERC (ca. 1895)

Pipes, 4ᵐᵉ Série *(Suite)*

Pipes, 8ᵐᵉ Série *(Suite)*

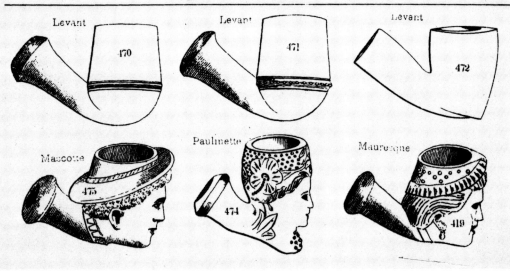

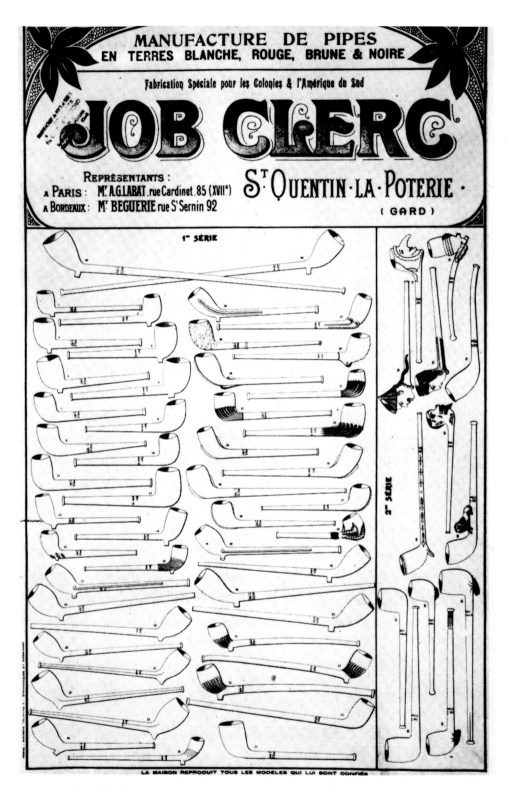

Two posters from Job Clere, manufacturers of clays in white, red, brown and black. Founded in 1812 at Saint-Quentin-la-Poterie, the firm closed its doors officially between 1969-1970. (Reproduction courtesy Jean-Léo)

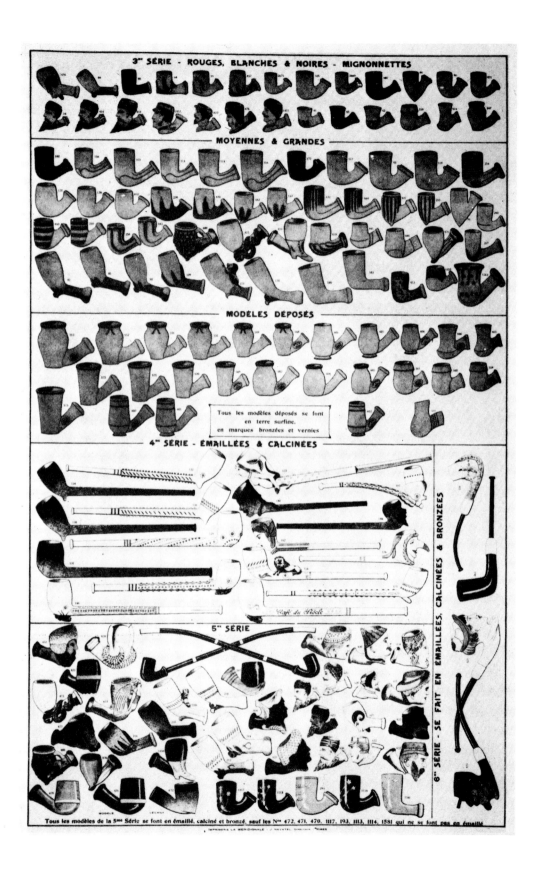

Hippolyte Léon Bonnaud of Marseille. (Reproduction courtesy Jean-Léo)

of Budapest. A potters' guild was established in 1680 in that town and by 1798, 138 master pipers were producing some ten to eleven million chimney-style hexagonal panelled red tone clay pipes per year, known as the Debrecen long clay.[14] In 1847 alone, ten million such bowls were exported. The other was the Nineteenth Century pipe genre of Schemnitz. North of Budapest, in a region once known as Old Slovakia, bounded by Austria in the southwest and Hungary in the southeast, the mining town of Schemnitz perfected a reddish brown or terracotta stoneware pipe bowl, often found in other hues of matte black, blue and green. This glazed and mottled substance had the texture of porcelain, the patina of aged meerschaum and the durity of stone. Schemnitz ateliers as Joh. Partsch and Z. K. Selmeczar manufactured some exceptionally ornate and filigreed pipes by a process which has never been analyzed, recorded or divulged. It remains a trade secret buried with Schemnitz in obscurity. The popularity of reds was eventually supplanted by the preferred whites and both Debrecen and Schemnitz acquiesced to the already established Western European versions. The distinctive Debrecen style of pipe is carried on today with the inexpensive, panelled and decorated white clay known as the Viennese coffeehouse. The Turkish chibouque, discussed in another chapter, is a member of the terracotta family still being made.

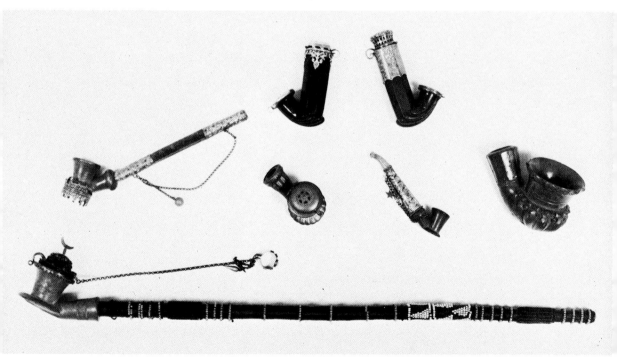

The Central European clays. At top center are two Schemnitz pipes, one in black with silver adornment, the other in terracotta with gilt filigree. All others are Turkish and Egyptian terracotta chibouques of varying sizes and ornamentation. $50-250.

F. W. Fairholt, an Englishman, Honorary Member of the Society of Antiquaries of Normandy, of Picardy and of Poitiers, was enamoured with French not British clays and declared: "The magnificent pipes of the French market are got up for the delectation of the foreigners with whom the capital abounds, and for the pipe-collector, a being who rides a hobby liable to become franticly (sic) extravagant."[15] Fairholt had the wisdom to recognize the appeal of clays to collectors, but at today's prices, there is insufficient proof that he was correct about the extravagance of their cost unless $5.00 to $15.00 is considered exhorbitant for yesterday's clay pipe found nowadays at flea markets and fairs! Ah, but for that price, the infinite variety of clays to accumulate—the mass produced and much copied Jacob, the very rare bust of Christ (Gambier No. 800), steam engines, butterflies, cannons, skeleton heads and bearded emirs, sultans, sheiks and pashas. The English Bottle Collector's Dictionary, 1976, published the following suggested prices:

	Bowls Only	Complete Pipes
Plain	5p	£1.00
Decorated	10p–£1.00	£2.00–£5.00
Pipe in the form of a head	£1.00–£3.00	£2.00–£15.00[16]

The rarest Gambier...Christ. (Courtesy Jean-Léo) $250-400.

A petite clay figural of a grenadier. (Courtesy le Musée du SEITA) $100-150.

Three whimsical clays: Extreme left and right are clay "nose" pipes. In the center, a tribute to Punch of Punch and Judy from the Belgian clay firm of Chockier. (Courtesy le Musée d'Intérêt National du Tobacco, Bergerac) $50-75.

A well smoked 10" clay. This "Boston Bean Pipe" was purchased in 1878 for $40.00. (Ken Erickson Collection) $50-75.

The subject of pottery pipes rightfully belongs to this chapter for pottery is nothing more than ceramic, earthen-or stoneware. While the English, the first to make clays, were surpassed by Continental rivals, they were second to none in pottery and stoneware and developed a family of decorative pottery pipes which are appealing. These pottery pipes are essentially of three classes: the puzzle or snake pipe, the convoluted, sometimes grotesque, but always colorful productions

Prattware puzzle pipes in various convolutions. The center piece is inscribed D. G. 1809. (Courtesy US Tobacco Company Museum) Top: $1000-1500, center: $1500-3000, bottom: $1500-3000.

These 18th-19th century English pottery pipes go beyond the imagined! (Courtesy US Tobacco Company Museum) $1000-3000.

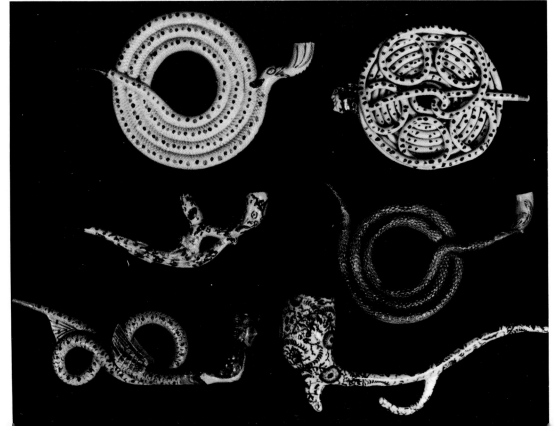

Examples of Staffordshire, Whieldon and Brampton ware. The free-standing oddities in the center row almost always portrayed pipe smokers. (Courtesy US Tobacco Company Museum) Top: $1000-2500, center: $1500-3000, bottom: $1500-3000.

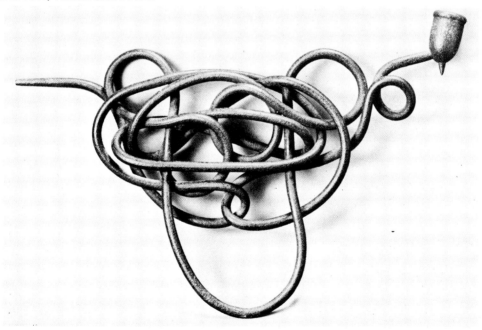

In French, it is faïence, but this 17th-18th century serpentine pottery pipe from the Rothschild Collection is less commonly encountered than the colorful and decorated counterparts from England. This may have been simply a continental expression, an early prototype or just a painting oversight on the part of those charged with quality control. (Courtesy le Musée de Grasse) $1200-2000.

from factories in Staffordshire, Bristol and Swansea—the familiar Prattware; Whieldonware, straight stemmed pottery pipes, sometimes with one, sometimes with many bowls emanating from a single stem; and, the glazed stoneware oddities from Brampton, Fulham and Nottingham. The inspiration for these imaginative pipes is unknown, but the vintage is between 1775–1850, supposedly the masterpieces of journeymen terminating their apprenticeship.

I conjecture that many English clay pipe-makers diversified at a time when demand for the clay smoking pipe was declining but this cannot be confirmed. It is a weak supposition since the signatures on a few puzzle pipes I have personally inspected bear no resemblance to those of clay pipe-makers; it is, however, a thought for further investigation. The greatest accumulation of such pipes was in the Bragge Collection and today, many are in the Willett and the W. D. and H. O. Wills Tobacco Collections, England, and the United States Tobacco Company's International Borkum Riff Collection. A Heide puzzle pipe in the PB 84 auction brought a respectable $375.00 for its three principal loops and surrounding subsidiary loops. Today, such a pipe is worth considerably more. Little more is known of this very characteristic English pipe from the end of the Eighteenth—mid-Nineteenth Century . . . they are exotic, extravagant, and very rare if found in superlative condition. They are unusual adjuncts to a "smokable" collection since pottery pipes are considered for decor, not for duty!

The clay was never the ideal smoking pipe . . . while it was portable, cheap and charming, it was temporary and brittle. It was popular among the working classes, the poor, fishermen, sailors. It was no match for briar and meerschaum, but in its own day, it was unsurpassed in variety and assortment, style and motif, decoration and ornamentation. For that reason, today's clay collector is able to amass a veritable infinity of fragments or complete pipes throughout his collecting lifetime and not catch one glimpse of a duplicate. He need only be on guard for reproductions.

"Think not of meerschaum is that bowl: away,
Ye fond enthusiasts! it is common clay,
By Milo stamped, perchance by Milo's hand,
And for a tizzy purchased in the Strand.
Famed are the clays of Inderwich, and fair,
the pipes of Fiolet from Saint Omer."
                    Burlesque Poem, 1853

Nineteenth century Whieldonware pottery pipes. (Courtesy Wills Collection of Tobacco Antiquities, Bristol, England) $75-175.

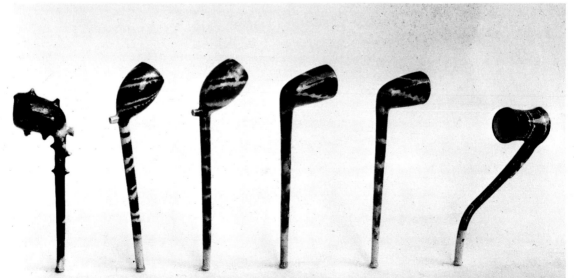

A 19th century polychromed pottery head of the devil with cherrywood push stem purchased in 1969 for $45.00. (Courtesy P. Naumoff Collection) $150-250.

FOOTNOTES: CHAPTER 2

2. Adrian Oswald, *Tobacco Pipes*, Connoisseur Concise Encyclopedia of Antiques, IV, 1959, p. 202.
3. Adrian Oswald, English Clay Tobacco Pipes (British Archaeological Association, London, 1967) p. 5.
4. Joseph Fume, A Paper: Of Tobacco (Chapman and Hall, London, 1839) p. 51.
5. Roger Fresco-Corbu, "Faces on French Clay Pipes", *Country Life*, June 14, 1962, pp. 1445-1446.
6. Ibid., p. 1446.
7. G. L. Apperson, The Social History of Smoking (G. P. Putnam's Sons, New York, 1916) p. 163.
8. W. A. Penn, The Soverane Herbe (E. P. Dutton & Co., New York, 1901) p. 156; Carl Adolf Vogel, Jagdschaetze im Schloss Fuschl (Droemer Knaur, Munich, 1974) p. 101.
9. Jean-Léo, Les Pipes en Terre Françaises du 17me Siècle à Nos Jours (Bruxelles, 1971) pp. 37-54.
10. Giusseppe Ramazzotti, "Classical Clays," *Pipe World*, Vol. 1, No. 0, Autumn 1969, p. 27.
11. Roger Fresco-Corbu, "The Rise and Fall of the Clay Pipe", *Country Life*, May 21, 1964, p. 1289.
12. Anon, Tobacco Whiffs for the Smoking Carriage (Mann Nephews, Cornhill 1874) p. 30.
13. Ibid., p. 31.
14. Vogel, *op. cit.*, p. 101.
15. F. W. Fairholt, Tobacco: Its History and Associations (Chapman and Hall, London, 1859) p. 180.
16. R. J. Flood, Clay Tobacco Pipes in Cambridgeshire (The Oleander Press, Cambridge, 1976) p. 47.

Wedgewood blue and white Jasperware pipe bowl fitted with a hide covered stem, circa 18th century. (Courtesy US Tobacco Company Museum) $350-800.

Chapter 3

THE PORCELAINS

"The great German china pipe is indeed and in truth a veritable sink of iniquity. There is no absorption, and it must frequently be emptied and cleansed, or it becomes a reservoir of disgusting poison."
    J. W. Cundall
    Pipes and Tobacco, 1901

The porcelain is, without question, the most controversial and paradoxical pipe for the ethnohistorian or collector. Generically, it is a material with much outward expression yet no one is prepared to believe that it could have ever been a quality smoking pipe. Porcelain is a strong, vitreous nonporous ceramic fired in ovens. It is on the one hand, as J. W. Cundall stated above and as a Veteran of Smokedom articulated: "The execrable china pipe is the mystery of the German. It has no absorption. It is a mere tobacco still, condensing the fetid juices in its reservoir, which must be frequently emptied and cleaned, or it is converted into a hubble-bubble of disgusting poison."[17] On the other hand, the porcelain pipe, to an experienced eye and to the tactile senses of an appreciative collector, is refined beauty, animated art. Thus, the porcelain remains a conundrum.

Wedgewood and Jasperware pipe bowls were made as early as 1781 but the earliest European porcelain centers were Germany, France and Italy. It was, moreover, the German who excelled most with porcelain as a medium for pipes. The earliest traceable German establishments were Meissen, near Dresden, 1708, Nymphenburg, near Munich, 1754 and Frankenthal, in the Palatinate, 1755 (?). Later factories were at Fulda, Berlin, Hoechst, Hohenberg, Augsburg, Limbach, Neudeck and Wallendorf. The celebrated rivals were Vincennes, France and Capo-di-Monte, Italy.

On January 23, 1710, the Royal Saxon Porcelain Manufacture was established at Meissen to capitalize on the ideas of Johann Friedrich Boettger, an alchemist, to whom the technique of Meissen is attributed and at the Leipzig Easter Fair of that year, the first porcelain pipe head was displayed.[18] As the Koenigliche Porzellan Manufaktur, KPM, the electoral arms of Saxony, crossed swords, became the Meissen hallmark in 1724; very good Eighteenth Century imitations have been found on La Courtelle, Paris, other later German porcelain and occasionally on British Worcester, Bow, Chelsea, Derby, Lowesoft and Bristol porcelain.[19] Among the many workers at Meissen, a certain Johann Gottlieb Ehder was chiefly in charge of the smaller works of art—the porcelain pipe relief heads of people, of real and imagined animals.[20] At a Nymphenburg factory, the prod-

A Wedgewood pipe bowl in blue Jasperware. On one side, a seated cherub reads a parchment scroll under a shade tree. The other side depicts three Grecian dancers. Purchased in 1967 for $150.00. (Charles P. Naumoff Collection) $500-800.

A Jasperware bowl with an ominous theme in black and white. (Courtesy le Musée du SEITA) $250-400.

ducts of Franz Anton Bustelli who worked there from 1756-1764 appeared in a sales catalog of that day and two of these pipe designs follow:

> "1. Modelled as a man's mask and with rococo scrolls and wave ornament in high relief, decorated in natural colours, pink and gold, mounted with metal-gilt lip, pierced cover and border chased with scrolls and foliage.
>
> "2. Modelled as a fore-arm and hand holding a cylindrical tankard decorated in colours with silver-gilt rim, cover with acorn finial and border."[21]

It was during the Eighteenth Century that at least one exotic style, or more appropriately, form and ornamentation appeared on pipes—Chinoiserie. Chinoiserie is defined as an ornamentation, characterized by intricate patterns and an extensive use of Chinese motifs. The porcelain pipe bowls which depicted these pseudo-Chinese scenes were tall, slender, most often hexagonal or octagonal panelled and were surmounted by highly ornate, filigreed, silver or gilt deckels or lids. Thus, Chinoiserie, at least in pipes, represented form as well as ornamentation! These pipes depicted imaginary scenes purporting to illustrate Chinese life and customs after travel books as Nieuhof's EMBASSY TO THE GRAND TARTAR CHAM, written in 1669, and were not a copy or adaptation of actual Chinese work.[22] Meissen was not the only factory to seize upon Chinoiserie. . . Augsburg was also noted for this style.

The Nineteenth Century expression of the porcelain represented the premier period of its popularity, the full blossom of its artistry, the period which should attract the collector most. Although the art of making porcelain was perfected around 1709-1710, it was in the Nineteenth Century that several developments concurrently influenced the porcelain pipe: it was fabricated and assembled as a four (sometimes counted as five) part pipe; there was the Biedermeier movement; the discovery of transfer painting; and, the regimental. In England, Worcester and Chelsea porcelain pipes prevailed. There was Sèvres in France, Majolica pottery in Italy. The Low Countries and Scandanavia were exposed to the porcelain movement, but the popularity was not of the same degree or magnitude as in Germany.

The first important influence was the gesteckpfeife meaning, figuratively, a pipe of many pieces. Around the end of the Seventeenth Century, an unidentified Austrian doctor (no doubt, Dr. Vicarius!) is said to have been responsible for the design of the four (or five) piece pipe: bowl, reservoir, stem, woven fabric flexible hose and mouthpiece. Gesteckpfeife was a generic term and was known by many names dependent upon whether the pipe head was porcelain, wood or meerschaum; and, at the peak of its popularity late in the Nineteenth Century, each occupation could call upon his style founded on distinctive ornamentation. One caustic description of gesteckpfeifen is too precious not to be quoted: "In the meantime let it be known that a German pipe consists of four parts: (1) the kopf, or bowl, to hold about a pound of the leaf; (2) the abguss, or lower receptacle, to catch the pernicious oil, which is afterward collected by religious societies and exported to the benighted heathen in Greenland; (3) the rohr, or stem, which is also used for educational purposes by German parents, and (4) the mundstueck, or mouthpiece, which is removed from the lips on rare occasions and when eating and drinking."[23] In Germany, the reservoir is also known as a suddersack or schwammdose.

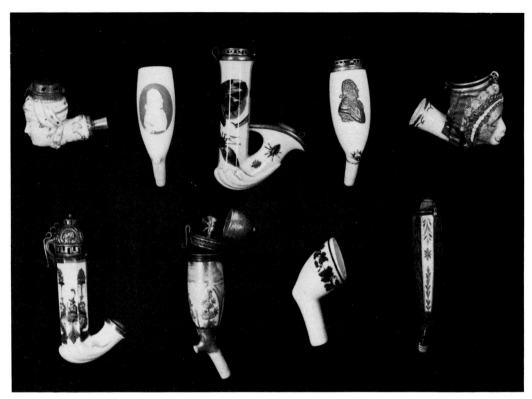

Porcelain peculiarities. Top, left to right: Meissen polychromed head; KPM porcelain with blue Jasperware style insert, silhouette of Washington; entomological porcelain with brass fitments; another KPM Washington silhouette in gilt; Meissen polychromed monkey with glass eyes. Bottom, left to right: Chinoiserie; a rare specimen of a double lid painted bowl of serpentine arches and flowers and the inscription: *aus Dankbarkeit,* two dancing children in silver, holding wreath inscribed: *Pour Toi;* clay shaped porcelain with circle of green leaves, marked KPM; slender paneled bowl of flowers and scrolls. $100-2500.

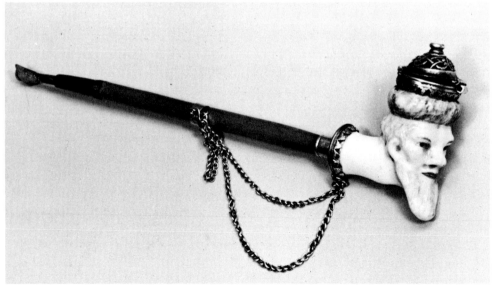

An 18th century polychromed porcelain pipe, crowned with silver cut-work lid, silver collar and retaining chain. (Courtesy H. F. and Ph. F. Reemtsma, Hamburg) $300-450.

A truly fine German porcelain pipe with horn abguss and finger holder. Reh and horn stem with porcelain insert painted en suite. A jagdstueck (hunting piece), 19th century. (Courtesy US Tobacco Company Museum) $250-500.

The "Y" shaped reservoir made the porcelain more palatable to the smoker who could either clean or replace it. It also gave the artist an opportunity to decorate the reservoir en suite with the bowl, although generally the reservoir was left undecorated. Stems varied in length and were precisely classified: quarter length, up to 12 inches; half length, 12 to 20 inches; three-quarter length, 20 to 31 inches; and long, greater than 31 inches. Stem materials included cherrywood, cedar, ebony, bamboo, bone, horn and antler, vulcanite and india-rubber.

The porcelain gesteckpfeife was a favorite of the Biedermeier era (1815-1840), adapting Gottlieb Biedermeier's art to the pipe, resulting in bowls painted with romantic to risqué scenes. Roger Fresco-Corbu, an English pipe collector who has studied them in depth, established twelve classes or groups in which just about any decorated porcelain would fit:

1. Portraits: men and women, the famous, the notorious, politicos, statesmen and artists.
2. Famous Paintings: copies of famous art in miniature, expressed in infinite detail and often bordered in gold to simulate a frame.
3. Military and Patriotic: historically important battles and campaigns and the military leaders which influenced them. This category includes the regimental pipe to be discussed later.
4. Student: university scenes—the duels, bierstube drinking scenes, the classical symbology of fencing equipment and fraternity caps.
5. The Chase and Sport: scenes of the field and stream, the hunt and other such pastoral activities.
6. Religious: biblical accounts and characters.
7. Fairy Tales and Legends
8. Domestic and Convivial: family reunions, gasthouses and courting couples, often classed as Defreggerbilder, after the artist Franz von Defregger.
9. Humor: satire, caricature, parody.
10. Trades and professions: guilds, occupations, associations.
11. Souvenir: panoramas of towns, resorts and national sites.
12. Ornamental: flora, fauna, birds, animals, Chinoiserie.[24]

So varied and diverse was the selection of quality painted porcelain ipes available to the common man that "... instead of a catalogue raisonné you may go to any pipe shop to know which are the best or at any rate the most popular pictures, by the miniature copies on the bowls."[25]

Another variety of porcelain bowl was the figural or character pipe, not to be confused with the earlier mentioned Meissen moulded heads. These included talons holding ostrich eggs, begging dogs, monkeys and peasants holding kegs or pitchers of beer. Since they were constructed to accept a stem directly without reservoir, they were never as successful as the elliptical shape porcelain bowl with reservoir.

The third development, transfer painting, was a change in manufacturing technique. With the passage of time, hand painted moulded pipe heads had given way to hand painting on plain white bowls, a technique which is easily identified by running a finger across or carefully inspecting the painted surface. Underglaze transfer painting in the middle of the Nineteenth Century eventually replaced hand painting and became the modus operandi for mass produced porcelains . . . a sacrifice of quality for quantity!

Generally, all bowls had a lid or wind cap—a deckel—and those found today without one were simply lost through time. Deckels were designed to fit snugly, blend in ornately with the theme of the bowl and were of silver, gilt, brass, chromium, nickel and other assorted metals. Again, with typical German precision, seven basic shapes were available:

1. Wuerzburger—an unornamented flat-top.
2. Hut—a hat. A slightly bevelled and unornamented lid.
3. Eckig—angular, cornered or edged. This type had open cut work around the edges.
4. Rost—fire-grate. This was a perforated or louvered top.
5. Imker—beekeeper. This style is not often seen. Surmounted on the deckel was an obelisk-shaped thumb lift.
6. Turm—a tower, spire or turret. A high domed deckel with finial.
7. Drahtdeckel—a wire screen. This deckel was a very early and popular style that mounted flush with the bowl rim, was removable and was fastened to the stem or reservoir with a chain.

The ultimate deckel appeared on the consummate porcelain late in the Nineteenth Century—the reservistenpfeife, the regimental, that last significant development in the life and times of the porcelain. The spiked helmet, pickelhaube, was to cap the very bowl which carried perhaps the most historical of Nineteeth Century messages, military service.

In Imperial Germany, after the Franco-Prussian War and prior to World War I, service in the military was rewarded with a parting gift of a beer stein, a regimental pipe and sometimes both. In all five branches of the Imperial Army, in some airship units and in a few Navy units, the custom was well established. The collector of regimental pipes is known to specialize by unit, garrison, branch or period. He can elect only bowls or intact complete pipes, colorfully ornamented with the Imperial red, white and black tassels, assorted accoutrements, unit and campaign lineage and other symbology. He can seek an even narrower subset since historically, there were various bowl styles. Early bowls generally portrayed the name of the reservist and his dates of service on one side; on the other side was a uniformed soldier and a civilian, a symbol of entry into or departure from the service. These early regimentals were surmounted with flat, unornamented lids. Around 1890, as the popularity spread, there was an entire market established for such memorabilia—cigarette cases, ash trays, steins and beer glasses. From 1900-1914, the apex of the reservist pipe, there were two basic bowl patterns: the one mentioned above, always smooth, flat, glossy and decorated with that early theme; the other was a raised laurel leaf, acorn and ribboned border surrounding portraits of soldiers, the Kaiser, the King or princelings. There were also many variations in the stem and its ornamentation. Testimonials and mottoes on these colorful, oversize bowls recall, movingly, the period of camaraderie, esprit-de-corps and unity of Imperial Germany's finest hours: furchtlos und treu—fearless and true; mitt Gott fuer Koenig und Vaterland—with God for King and Fatherland; Parole Heimat—watchword Homeland; Seig oder Tod—Victory or Death! A supersize

Super-size porcelains from western Europe. Extreme left: Danish regimental which reads: *Underkorporal Sorensen, No. 100, 1891.* Extreme right, another Danish with low relief birds and flowers, inscribed *Hans g. Jorgensen, No. 115, 1895.* In the center, a porcelain pipe with bowl and reservoir painted en suite. Upper left from Germany, an unusual shape in brown tones which translates: *Harmony is the most beautiful pillar in the temple of friendship.* Below, left, armorial shield with silver and red bars dexter, centered eagle, 3 helms, bear and lion rampant in gold, the arms of von Papen. $100-500.

A bust in white Parianware to the left, a female head in white porcelain to the right. (Courtesy le Musée du SEITA) $200-400.

The skull has traditionally been the subject of meerschaum carvers. Here is an expressive one from the 19th century in porcelain with an abguss moulded en suite. According to Arthur Machen, author of **Anatomy of Tobacco,** a skull's head pipe is to be smoked when in danger of falling in love or when being beguiled by "the monster woman." Purchased in 1968 for $65. (Charles P. Naumoff Collection) $300-500.

equivalent military pipe was made, at least, in Denmark. There may have been other Scandinavian countries which joined the movement to celebrate honorable military service, but none are known to this writer. In Germany today, the price range is between $150.00 to $400.00 for an intact regimental of average quality and average condition, when one can be found.

One infrequently encountered member of the porcelain family is Parianware, named so after Paros, a Greek island in the Aegean noted for its white marble. This fine, unglazed, hardpaste white porcelain was introduced around 1850 by both England and America, taking the form of biscuit figures and pipes. Parianware pipes were a short-lived novelty, fashioned as sculptured profiles and busts, dovetailed into long graceful wood, horn or ivory stemware. A pipe of Parianware is an attractive addition to the collector's miscellany.

The era of the porcelain had lasted some 200 years as an expression of the people. Sadly, during the first quarter of the Twentieth Century, political turmoil, global unrest and war contributed to the nadir of the German porcelain industry which had already transited from hand painting to transfer painting, from individual craftsmanship to mass production. Still, the porcelain is a desirable, fashionable and fascinating collectable. Prices remain nominal and supply, at present, still exceeds demand.

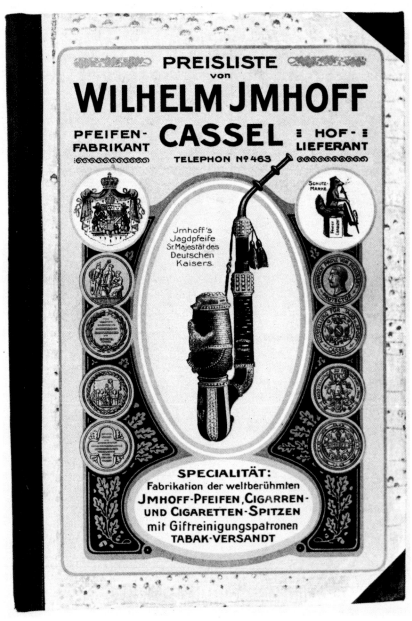

A pricelist from the Wilhelm Imhoff pipe company, Cassel, Germany, circa 1912. Pages 12 and 13 reflect just a few of the various porcelain styles and sizes offered by this factory just before World War I. (Author's collection)

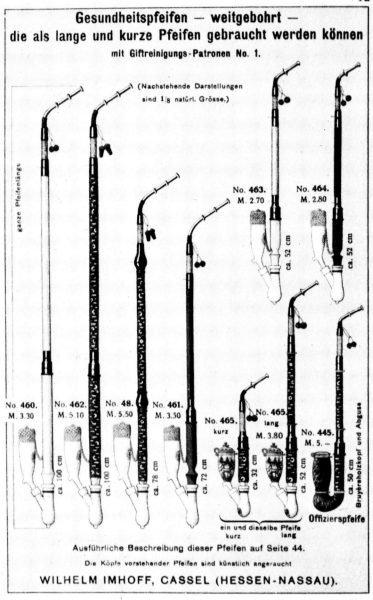

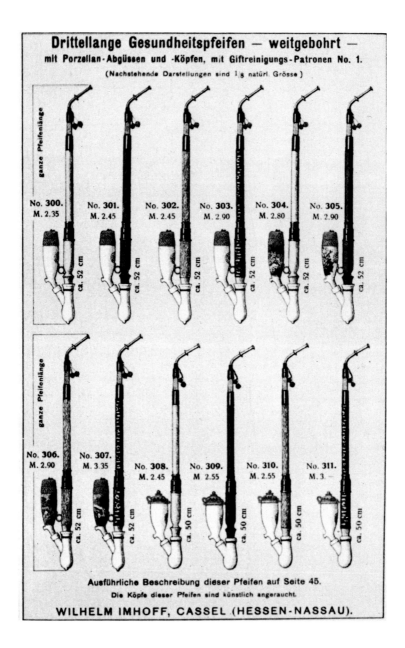

FOOTNOTES: CHAPTER 3

17. A Veteran of Smokedom, The Smoker's Guide, Philosopher and Friend (Hardwicke & Bogue, London, 1876) pp. 72–73.
18. W. B. Honey, Dresden China (Tudor Publishing Company, New York, 1946) p. 30.
19. Ibid., p. 169.
20. Ibid., p. 106.
21. Roger Fresco-Corbu, "German Porcelain Pipes," *Collectors Guide*, July 1972, p. 60.
22. Honey, *op. cit.*, pp. 65–66.
23. An Old Smoker, Tobacco Talk (The Nicot Publishing Company, Philadelphia, 1894) p. 86.
24. Roger Fresco-Corbu, "German Porcelain Pipes," *op. cit.*, pp. 61, 63.
25. Fairholt, *op. cit.*, p. 200.

Chapter 4

THE MEERSCHAUMS

"Well was it named écume de mer
The gracious earth so light and fair;
Mysterious cross of foam and clay,
From both it stole the best away;
If clay, 'tis such as sense might doubt of,
The same Jove made the Naiads out of,
If foam, then such as crowns the glow
Of beakers brimmed with Veuve Cliquot,
And here combined they sure must be
The birth of some enchanted sea,
Shaped to immortal form, the type
And very Venus of a pipe!"
From "To A Friend Who Sent Me A Meerschaum"
      by James Russell Lowell

Meerschaum has been called Venus of the Sea, White Goddess, sepiolite, sea-foam or froth, the Aristocrat of smoking substances! It has been identified as hydrous magnesium silicate, an opaque white-gray or cream colored mineral of the soapstone family. It has been written as $H_4Mg_2Si_3O_{10}$, $3S_1O_2MgO\text{-}2H_2O$, $2MgO_3 S_1O_2\text{-}14H_2O$ and even $Mg4(H_2O)3(OH)_2Si_6O_{15} \cdot 3H_2O$! Whatever its composition, meerschaum is not the bone of the sepia or cuttle fish, the residue or fusion of decomposed sea shells, petrified sea foam or a clay composition. It is a mineral! Although it has not been produced commercially in the United States since 1914, meerschaum has been found in Pennsylvania, South Carolina, Utah and New Mexico. It is mined in Nairobi but the quality is questionable when compared to that of Asia Minor, more precisely Anatolia, Turkey, where sufficient quantity and quality is found suitable for commercial use. Mined 30 to 450 feet below the surface of the earth, the magnesium content provides strength while the hydrogen and oxygen contribute to its porosity. It is, in the words of one, ". . . soft and light as a fleeting dream, creamy, delicate and sweet as the complexion of young maidenhood."[26]

The discovery and initial use of meerschaum as a substance for making pipes is shrouded in much mystery and myth. Even the origin of the word is obscure but the most generally accepted account is that a certain Hungarian nobleman, Count Andrassy, on an official mission to Turkey in 1723, received a lump of meerschaum from the Sultan (in some reports, the lump came from the Orient; in others, two lumps, not one). One elaboration on the Count details that he was an expert chess player. The Sultan challenged Andrassy and the latter lost three consecutive games. As tokens of esteem for the Count's subtle diplomacy, the Sultan gave him three gifts: a diamond encrusted dagger, two slave girls and a block of meerschaum. Only the meerschaum seems to have survived history and evidently left

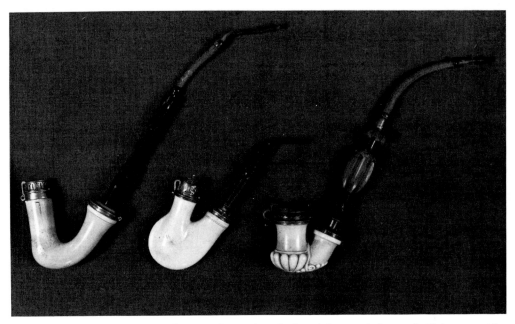

Three 19th century Austro-Hungarian meerschaum pipes. Left to right: a variation on the Debrecen, with long stemmed shank and silver deckel; an Ulmer in the center; and a Ragoczy with horn, wood and fabric stem, amber duck-billed mouthpiece. (Author's collection) $150-500.

Anatolia, if indeed any of this tale is true! Returning to Pesth, on the east bank of the Danube, he gave it to a cobbler who, when not making or mending shoes, made wood pipes; the Count was one of his patrons. This legendary cobbler (also said to have been an amateur or part-time construction worker) was Karol Kowates (also penned Karel or Karl, Kovacs or Kovacz). As the story goes, Kowates had carved two pipes; one pipe was given to the Count, the other was retained by the cobbler and that one allegedly is in the possession of the National Museum of Budapest. So begins the momentous success of meerschaum!

Another tale relates that when John Sobieski, King of Poland, rescued the beleagured city of Vienna from invading Turks, he had seen many Turkish articles, including pipes; the articles were made of luletaschi, a white mineral called pipestone believed to have been meerschaum.

And, there is yet a third story which, when published, should have rocked the collector world but it passed unnoticed. E. Reid Duncan claimed to have researched the Count at great length and could only find one Andrassy born in 1823 . . . one hundred years after Kowates was to have carved the first meerschaum.[27] Further, said Duncan, meerschaum pipes were being made before Kowates, even before Sobieski for it was a French artist, Louis Pierre Puget, student of the Italian sculptor Berini, who carved the first meerschaum in 1652. It seems that Puget was commissioned to carve two meerschaum horse statues to adorn the gardens of Château Vaux-le-Vicomte, then being built by Nicolas Fouquet, Superintendent of Finance to Louis XIV. Puget was paid a tidy sum of 20,000 livres for the two statues and was allowed to keep the meerschaum scraps from which he carved a pipe entitled Le Brigand au repos, the Bandit at rest, which "consists of four characters, the father, the mother, a small boy, a small

Bas-relief Kalmasch pipes, in meerschaum. 19th century. On the left, ostrich, egret and goose fleeing from hunter, mother-of-pearl inlaid ebony stem, fabric flex hose and horn mouthpiece; on the right, allegorical scene from the Middle Ages, horn and fabric flex hose and horn mouthpiece. (Author's collection) $150-450.

girl, two dogs, an eagle, and the trunk of a tree. The small statues are six inches in height and are in perfect detail. Each of the four heads are separated from each other and show expert execution even to the smallest detail."[28] Duncan elaborated that "the dimensions are 13 by 17 by 5 inches" and ". . . Puget, having finished the masterpiece, bestowed it upon his mistress. She, in turn, gave it later to her favorite André Le Maitre. Le Maitre, as generally happens to artists, having fallen into difficulties in indebtedness, was forced to sell the pipe to a pawn broker in Paris. Here it was redeemed by two citizens of Quebec, Mr. Emile Jacot, a jeweler, and Mr. J. B. Houde, a manufacturer."[29] Another commentary stated that with the scraps, Puget carved one pipe for himself and liked it so much that he carved another for Fouquet. Where did Puget obtain the meerschaum? Where is the Brigand au repos? Duncan's narration may be the earliest account of meerschaum, it may be a hoax. Meerschaum could have been used even earlier than 1652 but however devious, cryptic or circuitous the route, meerschaum came from the East and was popularized in the West.

Meerschaum's role from that moment forward is less doubtful. By 1745, the small village of Ruhla in Thueringia, Saxony, was a meerschaum center. In 1750, Kristof Treiss of Ruhla exhibited his meerschaum pipe bowls at a Leipzig fair.[30] By 1800, Ruhla's population had grown to 5,000 and 27 factories employing over 150 persons were occupied with making meerschaum pipes. Two of those Ruhla carvers, Johann Wolfgang Wagner and Jakob Hellman had received honorable mention for their pipe bowls exhibited at the 1776 Frankfurt industrial fair.[31] And, in 1794, a certain Michael Strassner from Pappenheim by Treuchtlingen in Bavaria exhibited his meerschaum wares—pipe bowls and meerschaum lined wood pipes—at an industrial fair in Ulm.[32]

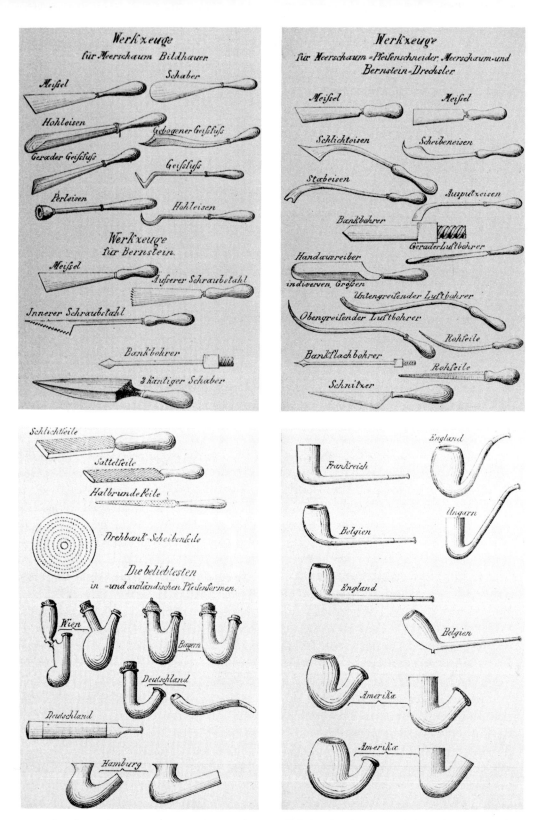

Drawings from **Die Meerschaum und Bernsteinwaren Fabrikation** by G. M. Raufer, Vienna, 1876. The top two illustrations are tools for meerschaum and amber carving. The bottom two are, in Raufer's opinion, the most beloved pipe styles. (Author's collection)

The latter half of the Eighteenth Century is replete with anecdotes and acknowledgements of meerschaum pipe carvers but their principal effort was bowls. Stems of horn, antler, ivory, ebony and cherrywood came from other parts of Europe where guilds had perfected such skills. Fitments and mountings, for example, were made by silversmiths such as Benjamin Barling, a London firm which chartered its own course in 1812 to carve meerschaum pipes; today, that firm is known as B. Barling & Sons, Limited.

In 1799, a mechanic in Erlangan, Germany, T. A. Thomas, composed a slight but convincing monograph on meerschaum: PRAKTISCHE ANLEITUNG MEERSCHAUMENE PFEIFENKOEPFE ZU VERFERTIGEN. This PRACTICAL GUIDE TO MAKING MEERSCHAUM PIPEBOWLS was an enthusiastic and brave effort to laud the superiority of meerschaum for pipes over any other material used to that time. Thomas unwittingly contributed greatly to the spread of its popularity.

Industrial expositions were not a new idea on the Continent and the British seized the opportunity in 1851 on a grand scale with an exhibition of industrial products from all nations. The Great Exhibition was so large and so lavish that it became the forerunner of subsequent World's Fairs. This London event, also known as the Crystal Palace Exhibition, had one section devoted to miscellaneous manufactures and small wares. The firm of Lux Brothers, Prussia, won honorable mention along with the firm of Louis Bolzau from Ruhla. A Barling bas-relief meerschaum bowl with ornate cherrywood stem and amber mouthpiece won an award. A silver-mounted meerschaum made by M. Held of Nuernberg, representing St. George and the Dragon, appeared in the illustrated catalogue published in conjunction with the exhibition. "We have selected for engraving this out of several drawings sent us by one of the most successful manufacturers of Germany. The article is one upon which much ingenuity is expended; it is often embellished with great skill and taste, and is not unfrequently made costly by the exercise of artistic talent; indeed, a very large proportion of the young Art of Germany is employed in modelling, carving, or decorating these meerschaums. In Germany there are few more productive articles of trade; they are exhibited in the gayest shops; and their ornamentation is generally expensive as well as beautiful."[33]

In 1850, the firm of Ganneval, Bondier, Donninger, GBD, set up to manufacture meerschaums in Paris.[34] That date should not go unnoticed since Cardon and Company claimed to be the first atelier in France to confection meerschaum pipes and that date was 1852.[35] During La Belle Epoque in Paris, it is said that there were 80 carvers, four per arrondissement, one per quarter on the average and two manufacturing establishments, Mathisse and GBD.

The famous meerschaum centers in Germany were Lemgo, Lippe-Detmold, famous for its horse carvings, the aforementioned Ruhla and Vienna where, without doubt, were found the master carvers from the middle of the Nineteenth Century to the Hitler regime. Without a slight to those craftsmen of Prague and Budapest, Milan and Bremen, the Viennese were truly unsurpassed! Remarkably carved feminine and mythological figure pipes have been attributed to Vienna where, around 1850, over 50 firms employing 10–15 carvers each, were actively engaged in producing pipes. In the last half of that century, there is evidence that

Ingenuity and elaboration were the forté of an Austrian artist, Moritz von Schwind, praised by Goethe for his fanciful and suggestive inventions. His *ALMANACH VON RADIERUNGEN,* Almanac of Sketches, 1844, contained six mystical, magical, even mythical designs for the meerschaum carver. The six are reproduced from a later edition of 42 sketches which appeared in a slim volume entitled *RAUCHGEBILDE-REBENBLAETTER,* Rotapfel Verlag, Zuerich, 1952. The first is in the form of a stove with a soup tureen lid.

Another for the imagination: a sultan awaiting his Turkish coffee.

This one is a gondola...with Orientals on board!   This one appears to be a winter pastoral scene.

Knights errant and castles.

Another country scene.

A Moritz von Schwind concept comes to life in this fanciful castle and moat configuration. (Martin Friedman Collection) $400-1200.

there was either an exodus from Vienna or proselytism or both. One London firm, Edwards and Company, brought in carvers from Vienna in 1862.[36] In the opinion of one Englishman, not only did Germans have the industry almost entirely in their own hands with Vienna as its headquarters, but in England "in the meerschaum-pipe manufactory among us the hands are entirely German, though some English lads are now being initiated into the work."[37]

Around 1880, large French firms as Maison Lancel solicited the best Austrian workers to live and carve in Paris, the hub of French meerschaum carving. Others came to Paris from Vienna of their own volition: S ɑmmer Brothers, M. Ziegler and Golstche whose grandson continues today as a third generation carver with a brother-in-law, H. G. Guyot, in the galleries of Palais Royal, Paris. The very masterful carvers, Reischenfeld, Nolze and Skopec of Vienna, were descendants of Austrian or Hungarian Jews. They disappeared, one by one, during the Reign of Terror of the 1930's and with them a skill unduplicated anywhere in the world today. M. Heimann was the last meerschaum sculptor to leave Vienna and he entered Paris in 1941.[38]

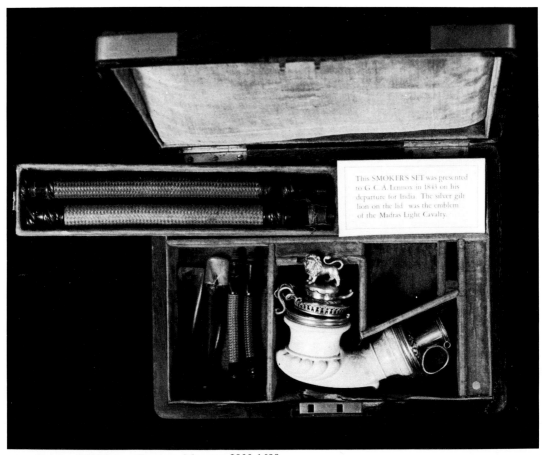

Courtesy US Tobacco Company Museum. $800-1600.

This bas-relief meerschaum bowl is from England, late 18th century. It is a classical Middle Ages battle scene. The figure surmounting the lid raises a question as to the theme—an Oriental or Middle Eastern potentate? A snake adorns the silver collar. (Courtesy le Musée du SEITA) $300-600.

    These were not, however, the only names which appear on the etuis of fine collector pieces. More frequently encountered are the trademarks, logos and imprints of BBB and H. Perkins, London; Bartolemo, Venice; L. Gambarini, Naples; Edoardo Flegel, Milan; Adler and Medetz of Budapest; Simenon of of Marienbad; Josef Dolezal, the Brothers M. and Emanuel Czapek of Prague; E. R. Rabe, Brussels; Georg Kopp, Dresden; Schnally, Bremen; the Brothers Franz and Carl Hiess, Ludwig Hartmann and Heinrich Schilling, Vienna; Arthur Schneider, Leipzig; Au Pacha, Dijon; D. Macropolo, Calcutta, Bombay and Colombo and, at least one woman, Helena Sofia Isberg, Smaland, Sweden.

    But, what of America during this period? Emigrants to America included many European meerschaum carvers. In 1855, F. J. Kaldenberg joined forces with an Armenian, Bedrossian, by name, who had brought two cases of raw meerschaum into the United States when he left Europe; the two set up a factory at the corner of William and Broad Streets in New York City, believed to be the first American meerschaum workshop.[39] It was not long after that when William Demuth & Company, the firm with the respected WDC triangle logo, was established. In 1862, a shop was opened in Manhattan; in 1896, Demuth opened a large factory in Richmond Hill, Queens, at Park Lane South and 101st Streets.[40]

Montebello, Palestro, Magenta, Marignan and Solferino were battles of Lombardia fought during the Franco-Austrian War which ended on June 24, 1859. Napoleon III, as victor, induced Maximilian, ex-Archduke of Austria and brother of Franz Josef, to become Emperor of Mexico. This carved pipe from J. Sommer, Passage des Princes, Paris, was given to Maximilian by Napoleon III as a token of friendship. Etched in spiral on the shank are those battles, the Napoleonic eagle and the motto, *Honneur et Patrie*. A credible audit trail has been established from June 19, 1867, the day on which Maximilian was executed. He gave the pipe to a Mexican guard outside his cell in gratitude for special favors accorded the imprisoned Emperor. That guard eventually married and heired it to his son who sold it to Fritz Stephan from Chicago who was visiting the Mexican state of Nuevo Leon. Stephan used the pipe as collateral for a $550 loan in December 1935. By 1958, it was owned by Iwan Ries Company of Chicago, appraised at $1,500 and was on display at the Chicago Public Library. In 1979, Mike Clark became the owner, having paid $3,000 for it at an Ohio gun show. For those who own the first edition of Dunhill's THE PIPE BOOK, Plate XXIV includes a sister pipe entitled Veteran of Crimea; its stem is a mirror image to that on this bearded gentleman, indicating the possibility that a series of Sommer meerschaums were made to commemorate the Franco-Austrian War. (Mike Clark Collection) $500-2000.

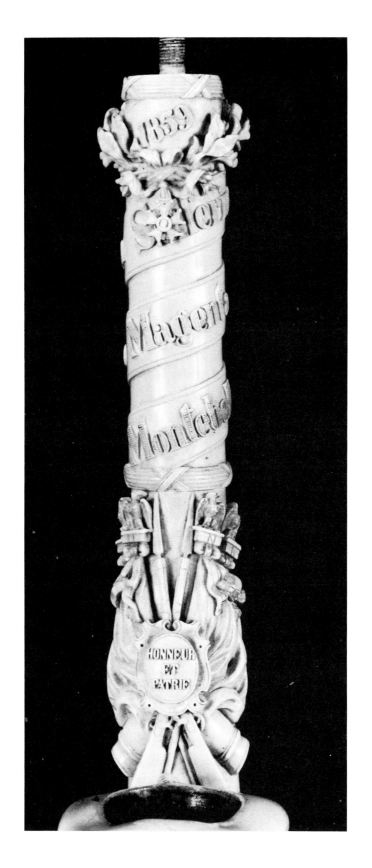

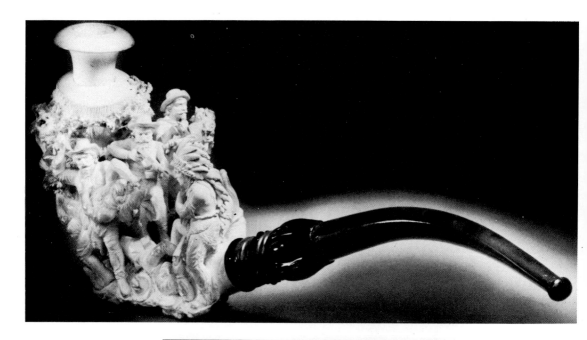

A New York City artist, Viennese born Carl Kutschera, owned a pipe shop on Third Avenue and carved this magnificent full and bas-relief meerschaum entitled *Custer's Last Stand*. Although not all the figures can be seen, there are eight men and a horse engaged in the battle of Little Bighorn where Custer died in 1876. The dimensions of this commemorative masterpiece are 17¼" x 7¼". This work of art was advertised for sale in antique journals during 1975-1976 for a firm price of $8,500. (Herbert G. Ratner Jr. Collection; photograph by Terry DeGlau) $2000-3500.

There were the numerous but unrelated Fischers in Massachusetts and New York. As to the former, Vienna born Gustave Fischer Sr. came to America in 1881, was a carver for 45 years, principally in the employ of E. P. Ehrlich Company of Boston and produced some of the most elaborate meerschaums known today. His son, Gustave Jr., followed in his father's footsteps, owned a shop on Massachusetts Avenue in Boston, followed by one in Jamaica Plain. Fischer Sr. passed away in the 1930's at age 89 after 50 years in business; the son died in 1975 at age 88, having been a pipe craftsman for almost 74 years.

This meerschaum pipe appeared on the cover of *Hobbies* magazine, October 1946, as an introduction to a feature story on the Heide Collection. It is an intricately carved pipe of naiads frolicking with Pan, after a painting by Adolphe William Bouguereau and is from Vienna circa 1875. It is 9" in length and, with its fitted case, sold for $700 in April 1878. (Copyright PB Eighty-Four, New York) $450-900.

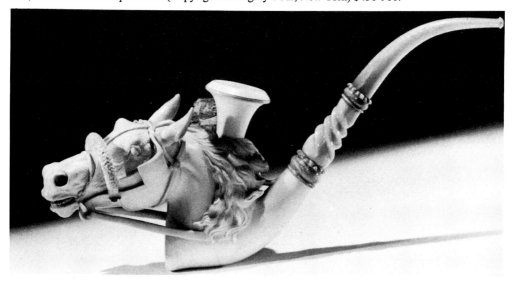

The famous trotter, Directum. This is a Gustave Fischer Sr. creation to honor Directum, the Trotter of the Year in 1918. As a six year old, Directum was the leading money winner with $13,217. The horse was owned by John W. Coggershall of Providence, Rhode Island, whose name appears on a silver plaque on the carrying case. (Courtesy US Tobacco Company Museum) $1200-2500.

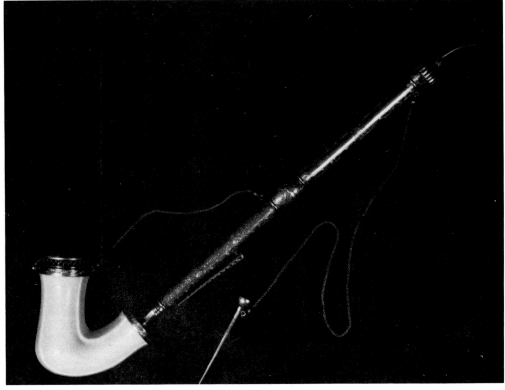

This meerschaum pipe was manufactured in Hartford, Connecticut by L. T. Welles Company. The shank is 18KG, highly chased and finished in a redolite mouthpiece. A retaining chain secures the bowl to the stem while another secures the pricker when removed from its sheath. This was a presentation piece to the purser of the Steamship Cortes. It now belongs to the US Tobacco Company Museum. $250-600.

Six of Gustave Fischer Sr.'s finest scuptured heads carved around 1900 and purchased from D. P. Ehrlich in 1978 by the US Tobacco Company for its museum. $300-700.

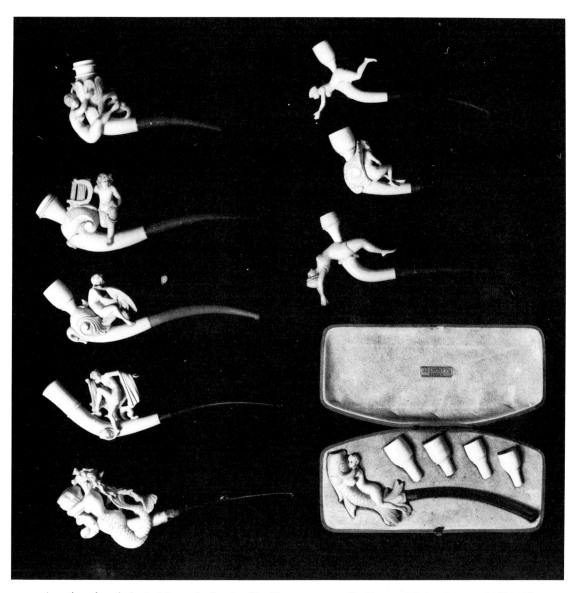

Aquatic and mythological finery for the cigarillo. They were carved by Gustave Fischer Sr. around 1900. All have fitted cases and additional bowls as the example in the lower right-hand corner. (Courtesy US Tobacco Company Museum) $250-550.

Until recently, Arthur C. Fischer, a sixth generation descendant of another carving family, lived in Orchard Park, New York. The earliest Fischer was appointed Royal Pipe-maker to the Prince of Saxony in 1742 and this tradition was carried on by the second and third generations. The fourth generation, August, came to the United States in 1867 and lived in Brooklyn where he and his sons Gustave and Otto were employed by W. Demuth & Company. August and his family lived for a while in Rochester, then relocated to Buffalo in 1892 and opened a shop on Main Street. August executed a significant number of important large-scale meerschaums for the 1901 Pan American Exposition in Buffalo. Gustave carried on the family tradition with the House of Fischer in Buffalo and diversified

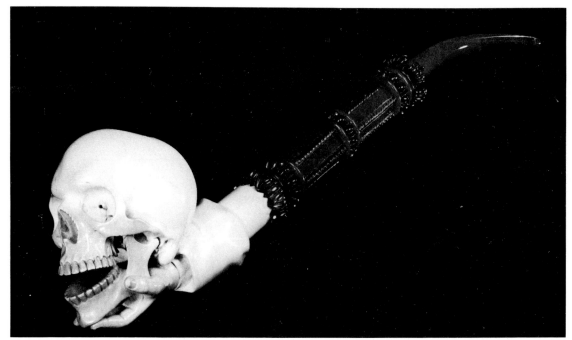

This is not just another skull pipe! Carefully inspect this August Fischer mammoth, carved for the Pan American Exposition in Buffalo, 1901. Dimension alone will separate this pipe from others: 30" in length and 7" in height. The hand-turned and hand-filed amber stem was crafted by Gustave Fischer, August's son. The precision and exactitude of this pipe qualify it for use in a medical lecture. Veins, finger-nails, cuticles and an open-mouthed skull make this a rarity. (Mike Clark Collection; photograph courtesy Arthur C. Fischer) $2000-3500.

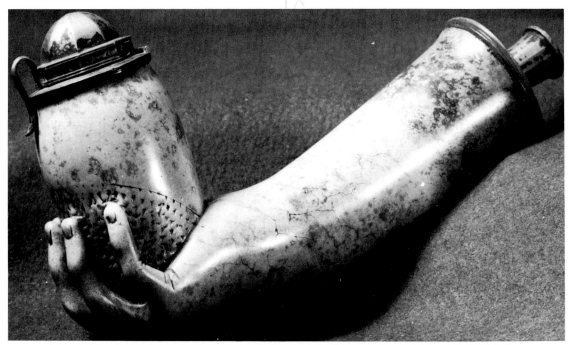

Hands holding cups or egg-shaped bowls are commonplace meerschaums but a whole forearm is an infrequent find. The length of this 19th century European pipe is 9" and height is 5½". (Ken Erickson Collection) $300-600.

into amber, briar pipes and artificial meerschaum coloring to order. Otto returned to New York City and his own shop. Arthur continued with briar pipes in Orchard Park, New York from 1956, while a relative, Paul of New York makes the House of Fischer meerschaums. On June 30, 1979, Arthur C. Fischer wrote the last chapter in this 250 year old tradition with his retirement to Florida.

I now return to the pipe. As to dating, there are a few universally accepted rules. Until 1850 or so, pipe bowls were generally large and came in an assortment of plain, bas-relief carved and sculptured. Ornamentation consisted essentially of mountings—lids, collars and shank trim in brass, silver and gilt. Shanks were short since these were fit with long push stems of assorted materials. The late Nineteenth Century witnessed a transformation from this type to the intricately carved meerschaum pipes. There was a reduction in bowl size, a lengthening of the shank, the expanded use of amber and the introduction of even smaller mirror-image pipe styles as the cigar, cheroot and cigarette holder to meet the demands of the cigar and cigarette smoking public. Gradually, as carved pipes became the order of the day, figural heads of personalities and other quaint subject matter appeared. As Fresco-Corbu indicates, although the blackamoor, the nubian, the Negro was used on tobacco statuary as a symbol of the trade in the Seventeenth Century, his appearance on a pipe bowl, with plantation straw hat, must be dated after the celebrated UNCLE TOM'S CABIN in 1851. The infinite variety of female busts with broad brimmed hats and finger-curl hairdos are easily traced to fashion magazines of the day and books on the history of costume . . . consider the time frame of the idealized American girl of the 1890's represented by artist Charles Dana Gibson and the Dolly Varden costumes after a character in Dicken's BARNABY RUDGE. Personage pipes were meant to be exact representations, so determine the approximate age of the personality portrayed on the pipe and any Who's Who will tell you the rest! Myriad pipes remain unexplained and perhaps the smoking case or carrying etui may provide the only clues.

Nineteenth century meerschaum wares: for the cigar, cigarillo, cheroot and cigarette. (Courtesy Wills Collection of Tobacco Antiquities, Bristol, England) $75-800.

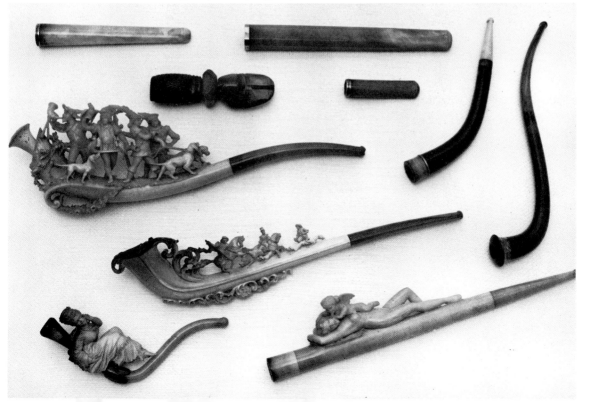

A young girl in garlands carved in Austria for the Vienna World Exposition, 1873. (Courtesy Austria Tabakwerke, Vienna) $300-600.

A classic pose of the last half of the 19th century. Very Victorian and very well carved. (Courtesy Wills Collection of Tobacco Antiquities, Bristol, England) $250-400.

This was the attention getter at the April 1978 PB 84 auction! Lot 178, a masterwork of Arthur Schneider. Leipzei, is a giant cheroot holder with three high relief Imperial cavalry soldiers on horseback, with armorial and bezelled ovoid amber ornamentation. This piece measures 18" in length and is late 19th century. Final price at the auction—$3,400. (Mike Clark Collection: photograph copyright PB Eighty-Four, New York.) $2500-4000.

There is no accepted date at which push stems fell into disuse. In fact, many early Twentieth Century pipe bowls were made to fit such. Horn and bone were the traditional industry stems for this newer style of meerschaum pipe, but it was amber which rapidly supplanted both. As to exactly when and how the marriage of meerschaum and amber occurred, no one is really sure but it was a natural and logical one. Amber is the fossilized resin of pine, known to the Greeks as elektron for its ability to retain static electricity. It is a gemstone, the lightest of gemstones and varies from colorless to black. The most popular colors are transparent golden yellow, cloudy buttercup, opaque and antique brown. Amber is known by other names: beeswax, wax stone, Baltic, Burmese and root amber. It is a substance hard enough to resist teeth marks; it retains a smooth surface while exposed to saliva; it does not transmit heat rapidly, burn the lips, crack or splinter. It is tough, lightweight, beautiful, easy to clean, pungent yet aromatic, tasteless—so why not?

Meerschaum bowl depicting Father Rhine and Lorelei as river God and Goddess. Between them is carved the Hapsburg coat of arms. This pipe was a present to a member of the Russian Imperial family. (Courtesy Douwe Egberts' Pipe Room, Utrecht, Holland) $350-900.

Sadly, as in so many things original and beautiful, there is the problem of reproduction, fake, forgery and substitute. The meerschaum pipe and its amber stem were not exempt. Vienna, while noted for its quality block meerschaum pipes was also the center for massa, pressed, false or sham meerschaum, often identified as Wiener or Vienna meerschaum; Vienna made no meerschaum, only meerschaum pipes! Cardon attributes unequivocal superiority to the Germans in imitation meerschaum.[41] There were many compositions and there was no true or valid test for false meerschaum. "Burnt gypsum slaked with lime or a solution of gum-arabic forms a hard plaster which, smoothed and polished with oil, assumes a marble-like quality. A very hard imitation of meerschaum is also made of hardened plaster of Paris, polished and tinted with a solution of gamboge and dragon's-blood, afterwards treating it with paraffin-oil or stearic acid."[42] "It should be added that the chips and dust resulting from the working of genuine meerschaum are bonded together with a solution and moulded and this is also sold as meerschaum."[43] "The waste from these carvings is ground to a very fine powder, and then boiled with linseed oil and alum. When this mixture has sufficient cohesion, it is cast in molds and carefully dried and carved, as if these blocks of mineral had been natural."[44] One enterprising manufacturer concocted a recipe which called for peeling potatoes and soaking them for 36 hours in a solution of water and sulphuric acid. The resultant mass was dried on plates of chalk or plaster of Paris, in hot sand, for several days at which time the compound was ready for the carving of meerschaum pipes. "Mock meerschaums are sold in London at various prices, from half-a-crown to half-a-guinea. They are chiefly bought by persons who have no knowledge of the real article. Pipes of this kind always colour in the wrong place; the upper part of the head, which in a genuine well-smoked meerschaum is white, in a short time becomes spotted with a dirty brown; while the part between the culotte and the socket seldom becomes colored at all, except in irregular patches of a dirty yellow."[45] At half the price of real, it was estimated that about twice as many false meerschaums as real were manufactured per annum after the processes were mastered.

One pressed meerschaum, for example, is the shepherd pipe. Typically, it depicted a pastoral scene, a fleeting deer, a horse, hunter, gamesman, warrior. Such pipes were richly and evenly pre-colored a mature brown at the factory and almost always had a false date engraved, incused or incised into the bowl—very early dates for meerschaum, I might add as 1710, 1730. Another indication of a false meerschaum is a mottled or spotty finish on a pipe in mint condition or one only slightly smoked. That is indicative of an uneven or poor quality moulding process in which time, temperature changes and exposure to sunlight have taken their toll; the wax has dissipated, leaving the appearance of spots and flecks. However, be not deceived! Some false meerschaums were so superior in craftsmanship that only by smoking it could the discerning eye recognize massa by the patchy ivory to brown hues throughout the pipe with an almost immediate change in color at the top of the bowl when in the real thing, that part normally, generally, usually, almost always turns color least. Weight may be a test, since lightness is an expectation in real meerschaum but specific gravity varies with the quality of meerschaum.

In the Great Exhibition of 1851, there was a display of composition bowls and cigar holders. A report of the exhibit stated that composition "bowls are distin-

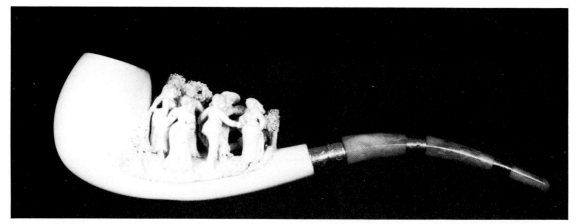

A handsome example of high relief carving on the shank of this 12" meerschaum from D. Macroplo, Calcutta, Bombay & Colombo. Five playful girls entice a young man to play blind man's bluff. (Irving Landerman Collection) $400-800.

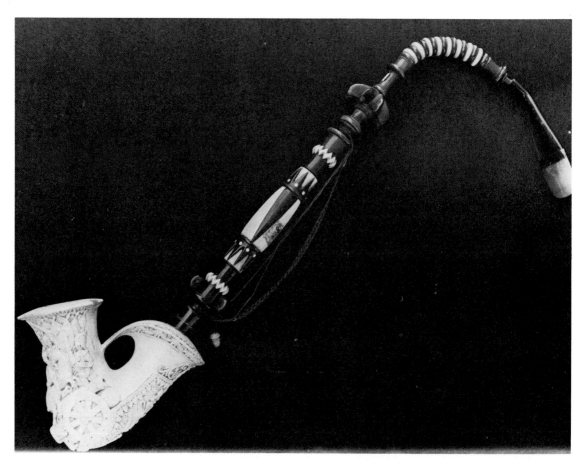

A 19th century Hungarian commemorative pipe to celebrate the defeat of the Saracens in the Thirteenth Century Crusades. The bowl was purchased in Vancouver in 1977; the stem came from the Heide Collection, elaborate Italian horn and ivory geometric designs terminating in duck-billed amber. (Martin Friedman Collection) $400-900.

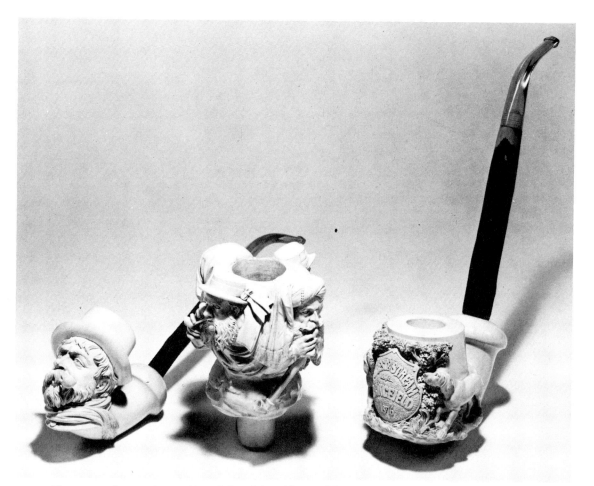

Three unusual meerschaums. From left to right: a hirsute Irishman smoking a cutty, cherrywood stem and amber mouthpiece. 11" in length. F. J. Kaldenberg Company, New York, purchased in 1978 for $250 (Author's collection); a Western interpretation of the smoking bowl for a nargileh or kallian, with four relief carved heads in the round: a beaded soldier, a bearded Arab, a Turk with native headdress and a gentleman with mutton chops, each smoking a pipe, 5" in height, purchased in 1978 for $550 (Martin Friedman Collection); a Kaldenberg pipe with shield, inscribed **Forest and Stream, Springfield, 1876,** with foliate background, flanked by a setter and a hound, 11½" in length, purchased in 1978 for $275. (Copyright PB Eighty-Four, New York) $300-600.

guished from real meerschaum by their greater specific gravity; but there is no very certain test . . ." and suggested an unconvincing negative test: "the composition bowls never exhibit those little blemishes which result from the presence of foreign bodies in the natural meerschaum; therefore, if a blemish occurs in a meerschaum bowl, which is very often the case, the genuineness of the bowl is rendered most probable; but as blemishes do not show until after a bowl has been used for sometime, the test is not of much value."[46]

In the United States, at the turn of this century, Montgomery Ward and Sears Roebuck catalogs offered mail order smokers' articles. There were shepherds in each company catalog, advertised as a German antique pipe: an imitation colored meerschaum bowl engraved in different designs, a nickel plated cover, an eight inch cherry stem with horn mouthpiece at the very low price of $1.40! Gen-

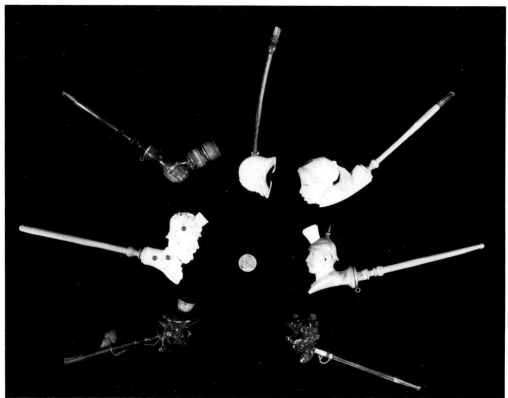

European sub-size holders in meerschaum. From 12 o'clock: pickelhaube with amber finial; nubian in tasseled cap, glass eyes; Kaiser Wilhelm with metal pickelhaube finial; a matched set of French pre-smoked delicacies; a helmeted soldier with amber beading and finial; French courting pipe. (Author's collection) $100-250.

A twin horse head meerschaum pipe in bypass. This regal pipe was carved for the Chicago Exposition of 1893, is now the property of Maurice Leonard and adorns his letterhead stationery. (Courtesy Maurice Leonard) $600-1000.

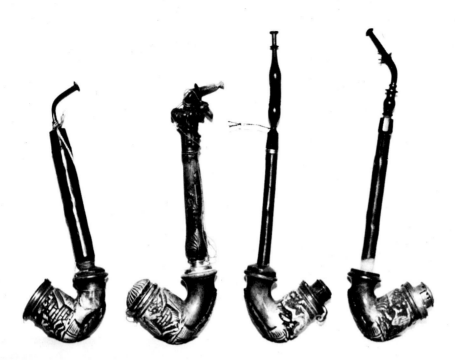

Four examples of the shepherd imitation or pressed meerschaum pipe bowl from the late nineteenth to early twentieth century. They are typically pre-dated and carry a bucolic or pastoral theme. (The late Hermann Thiede Collection) $50-100.

Chibouque style meerschaum with silver filigree spanning bowl to shank collar. Wood stem and amber mouthpiece. Rooster may have symbolized a product of France destined for Turkey. (Metro-media Collection, Courtesy Duke University Art Museum, Durham, NC) $100-200.

uine block meerschaums and imitation, chip or Vienna meerschaum pipes and cigar holders were also offered.

Amber's rising popularity also prompted substitution and occasional false advertising. False amber is amberoid, ambroid or ambroide, often used synonymously but incorrectly with celluloid. Ambrosine, however, is a yellowish to clove-brown variety of amber containing a high content of succinic acid. Ambroid is the nearest thing to real amber—it is pressed amber composed of actual particles, a rather misty looking composition. It is stronger than amber but does not improve with use as does resin amber. To compound matters, there were other pseudo-ambers used in the industry, though several were introduced in the early Twentieth Century: glassite, a light colored resin with a sticky texture; gedanite; kauri gum; copal; glass; bakelite, a compound of phenol or cresol and formaldehyde, containing no amber and no odor of pine. Then there was celluloid, condemned as highly flammable, offensive to taste and smell and causing impotence.[47] GBD claimed that while "pipes with ambroide were sold at exhorbitant prices, the public was being cheated, as the intrinsic value of the ambroide stems was only 30 centimes (3p) apiece."[48]

Most tests for imitation are precise but painful and costly. Amber melts at 536°F., for those who care to test suspect mouthpieces and watch the imitation reduced to liquid at lower temperatures. If amber is rubbed with ether, there is no effect; ether on other resins will act as a solvent and imitations will become tacky. Cutting amber with a knife will result in breakage, splintering and chipping; imitations will peel and can be cut. Two more simple but not foolproof tests may provide a confidence—amber will float in salt water and rubbing will produce static electricity. I caution that some imitations also generate static electricity! To avoid misinterpretation, my elaboration on amber was not intended to demean the imitations, since they were popular and acceptable in their day. Today, ambroid would be a distinct improvement over the synthetics which simulate even less quality, clarity and color than those latter day imitations.

No amount of dialogue could adequately and amply account for the innumerable pipe styles and types fashioned in the 150 odd year reign of the meerschaum until the practical, cheaper, more portable and less fragile briar came along. Instead, I have chosen a few romantic passages which I feel will convey a more precise message to the reader. To those who believe that antique meerschaum prices are sky-high today, here is a meerschaum to behold: "A gentleman drew forth from his pocket a short pipe, which screwed together in three divisions, and of which the upper part of the bowl—made in the fashion of a black-a-moor's head—near the aperture, was composed of diamonds of great lustre and value. Upon inquiry, I found that this pipe was worth about £1000."[49] That was 1839! In the words of another: "Meerschaums are frequently mounted in silver, and have sometimes been decorated with jewels, so that their cost has been excessive. They are generally enriched with ornaments in high relief, executed with much beauty and embracing a variety of design. The care with which this material may be moulded and fashioned by the artist (for such he is), who decorates the bowl, allows the greatest ingenuity and elaboration of design to be exhibited in this branch of art-manufacture. Most pipe-sellers and tobacconists can exhibit specimens which are perfect miracles of patience and labour, and are worth forty or fifty pounds each."[50]

One of the young men who brings pleasure to Bacchus...a young, pipe—smoking, vintager, late 19th century. (Courtesy le Musée du SEITA) $150-250.

A handsome cigarillo holder of the Three Graces. (Courtesy H. F. and Ph. F. Reemtsma, Hamburg) $200-400.

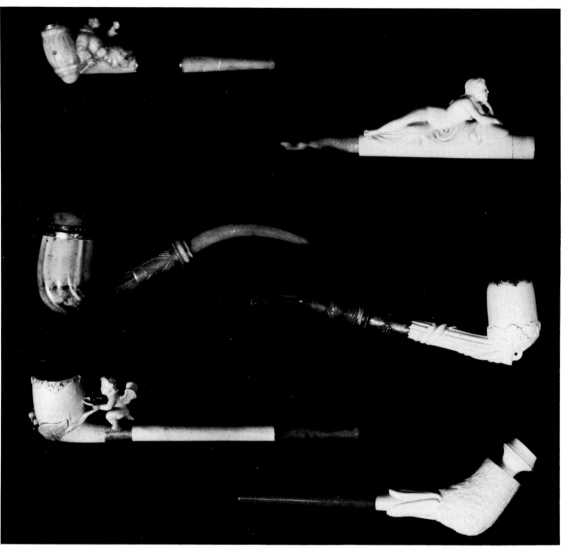

A melange of classic sculptured meerschaums from the US Tobacco Company Museum which express variations in style and theme from the latter half of the 19th century. $150-350.

Florentine arches and columns abound on this bas-relief bowl; rococo scrolls, turquoise, coral and rubies enhance the ultimate personification of Viennese carving from the 19th century. The lid is in meerschaum and at the thumb lift is the Austro-Hungarian double eagle. (Courtesy Austria Tabakwerke, Vienna) $2500-5000.

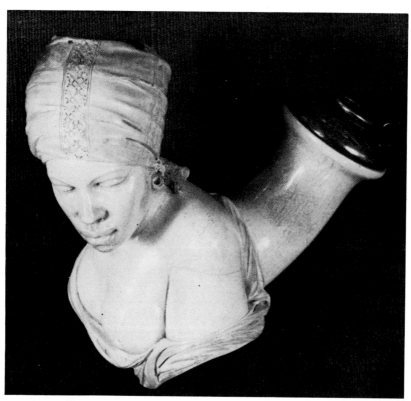

Giant nubian woman with one breast uncovered, measuring 6" x 6". Purchased in 1977 for $600. (Author's collection) $700-1200.

A full relief meerschaum of an imagined lion hunt in America. It is imagined because the headdress is distinctively American Indian but the facial features are Mongoloid. (Courtesy Austria Tabakwerke, Vienna) $800-2500.

Through the good offices of a fellow collector, Ken Erickson, selected pages of a rare manufacturer's catalog have been included. The 96 page Linger Company catalog is undated, but I conclude from the pipe styles and the appearance of Edison's incandescent light bulb in photographs of the workshop, that this catalog was printed more than 80 years ago.

Illustrated Catalogue (Wholesale & Export only)
B. Linger. Vienna, Austria and London
PREFACE

In issuing my catalogue my object is to enable my customers to order goods without needing sample collections. My catalogue shows all pipes in their natural size and colour, thus in most instances trouble and expense can be saved.

As all my present customers are thoroughly familiar with the quality of my goods, I have only to notify prospective customers, that I can send samples should they wish them; all that is necessary will be to mention to what amount the sample collection is to run.

Samples shown in this catalogue will be executed in first quality, throughout Briars, Ambroid, Meerschaum, Vulcanite and Cases. Second grade goods will only be made, when so ordered.

Where not otherwise mentioned all pipe cases will be covered in morocco, in black, brown, red and green. Colours as desired.

The lining of all cases will be of the best English plush, red, blue, green or any other shade ordered.

On pages 9 to 14 cuts of cases will be found made of various kinds of leather. Customers will thus be able to select according to their taste.

All my prices are quoted for pipes in morocco cases, as shown on page 60. If other cases are desired, the price will vary according to the prices of cases as quoted.

I would especially call attention to this fact, all my silver and gold mounted pipes are Hall marked and are punched H. F.

If desired pipes can also be mounted in electroplate.

My trade mark B. L. V. will only be put on pipes and cases at the express wish of customers.

All Briars can be had in natural colour, black or brown, the ambroid mouthpieces clear, cloudy or coloured. Any pattern can be supplied with Vulcanite or Horn mouthpieces, if desired, as shown on pages 61 and 62. Solid block Amber mouthpieces can be supplied in lieu of Ambroid. Prices for these on application. All Meerschaum samples can also be supplied in Imitation Meerschaum.

I make a specialty of Companions and do not hesitate to say, that these goods cannot be surpassed by any other firm upon the continent.

Pipes and cases can be stamped according to the wishes of customers with initials and any other marks. It will be sufficient to send a small sketch of the style of stamp desired with wording or letters.

In ordering goods, it will only be necessary to send the number of the cut, mentioning the colour of bowls, cases, leather and lining wanted.

Should anything be required, not shown in this catalogue, please write for it. For example, I have over 500 kinds of companions, but owing to the limited space of this catalogue, only 65 are shown. The same applies to pipes and tubes. Space does not permit showing the many other styles and shapes, which can be supplied.

The time for delivery of orders, will depend on the extent of same, but in all cases, where a special date is given for delivery, the utmost to carry out the wishes of customers is unvariably done.

TERMS on application.

INSURANCE—Goods for Export are insured by me and charged to consignee, unless other instructions are given.

PACKING: Cases charged cost price and unless otherwise ordered goods for export are packed in tin-lined cases.

DELIVERY f.o.b. London or Hamburg.

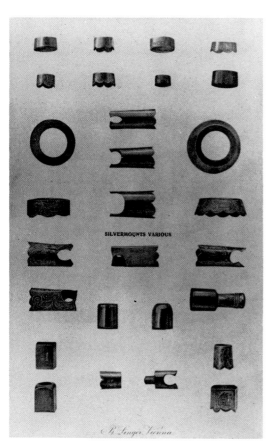
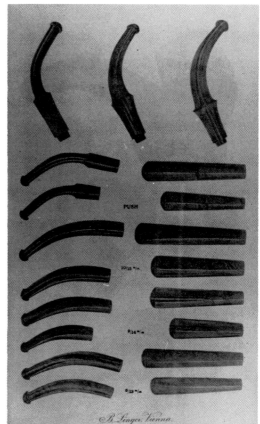
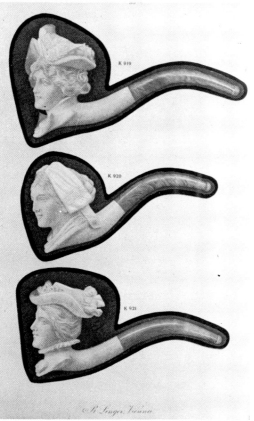
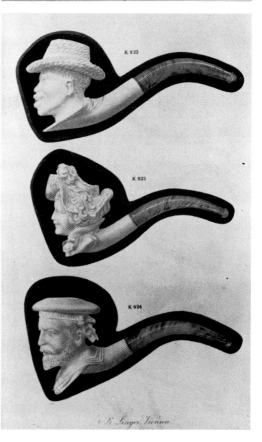

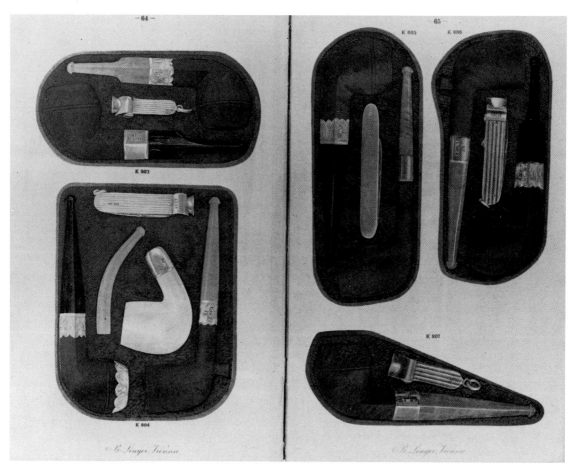

I save the pièce de résistance until last! Twelve years after the Kaldenberg Company formed, it participated in the Paris Exhibition of 1867 with a high quality display of American made meerschaum pipes. The success was so overwhelming that the Company exhibited at least once more, in 1876. Tinder Box International, Santa Monica, California, owns a pastel lithograph of a Kaldenberg meerschaum masterpiece exhibited at the 1876 International Exhibition and has granted permission for the narrative which accompanied that art treasure to appear in this book. Several paragraphs recapitulate aforementioned material but the passage is quoted verbatim to capture a fuller effect of the tempo and tenor of a fledgling American firm among the European meerschaum moguls more than a century ago:

> That pipes should be an especial feature of manufacture in the land which originally produced the weed they are destined to consume, seems natural and proper, but until within the last few years the reverse has been the case. Germany has almost entirely supplied the commerce of the world with the various varieties of the better classes of this article, from the cheap porcelain to the most expensive meerschaum. It is true, each land has its specialty in pipes, but England and France never produced any except the 'cuddy' or 'churchwarden' until within a recent period, when the latter country commenced furnishing the briar-root. This 'cuddy' pipe is made from a clay pe-

culiar to these countries, heretofore but seldom found in America. It can be baked hard, and remains at the same time porous, permitting to a certain degree the absorption of the juices of the tobacco. These pipes, invented shortly after the introduction of the weed into Europe, still retain the favor of the poorer classes of Europe as well as America. They are the pipes of the smokers of strong tobaccos. In Germany, on the contrary, the pipe most desired is one by which the enjoyment shall be lengthened while the quality of tobacco to be smoked is mild. In England and France, tobacco is moistened before the pipe is filled. In Germany, it is carefully dried. The well-known long German pipe, with its water bowl, necessitated by the quantity of tobacco stored in the pipe, allowing the saliva to mingle with the tobacco juices and settle at the bottom, has never obtained favor in America. The heavy Virginia and Maryland tobaccos are not smoked so long at a time, and a smaller bowl is desirable. The bowl, however, must be larger than the English or French, and the German makers have supplied this demand until within recent times. In Virginia, a pipe has at all times been made called the 'Powhatan;' it is of a clay which bakes red, and smoked through a long reed, it has excellences appreciated by smokers. It is, however, very fragile, easily broken, and does not absorb the essential oils to any great degree. For the supply of the want of a small compact pipe, which can be carried in the pocket and is not liable to break, Germany long had a monopoly. China, wood, and especially meerschaum pipes, were exported over the whole world, and for quite a period were articles of considerable revenue. The recent civil war, however, created a new demand, and at present briar and laurel-root pipes are manufactured in large quantities in the United States. These pipes have the advantages of compactness and strength, but are liable to burn when heated, and besides not absorbing the oil. For a long time the meerschaum has been the ideal pipe of both continents. It is absorbent and equally desirable for the strong as well as weak tobaccos. It can be used by the German for his light, highly-dried weed; by the Englishman or Frenchman for his strong, heavy composition; or by the Americans for their delicately-flavored tobaccos of Virginia.

Meerschaum is a clay found in Asia Minor, in pits or mines, near marshes. It is dug out in blocks, cleansed from a surrounding mould, and then packed on camels to the nearest port. The meerschaum (sea-foam), from its being so light, is composed of magnesia and silica or flint. It is more absorbent than any other known clay, and being so light and not easily broken, is eagerly sought after by all smokers. Doubtless from its contiguity to Turkey, which country controls the meerschaum mines, Vienna has long monopolized the manufacture of the meerschaum. The great factories there have obtained fame, fortune and reputation, and the possession of a real Vienna meerschaum has become an ambition with all lovers of the gentle weed.

When, therefore, Mr. F. J. Kaldenberg, of New York, exhibited a case of American-made meerschaum pipes at the Paris Exposition of 1867, where he took a prize so near the acknowledged centre of this manufacture, his enterprise excited considerable comment among experts. A German by birth, coming to the United States when but a child, he learned his trade here, and now vies with the best known Vienna houses, in the completeness of his manufacture and the beauty and finish of his articles. He has succeeded in creating an American industry which equals the best of Europe. With the usual energy of American manufacturers he has introduced labor-saving machinery, and imports his crude material direct from the markets to which it is sent for export.

In the Centennial Exhibition, Mr. Kaldenberg exhibited cases containing some fifteen hundred different pipes. The plate represents perhaps the largest meerschaum pipe ever made. It is twenty-eight inches high and eighteen inches wide. The large figure crowning the top is 'Columbia,' with the attributes of power, justice and liberty. The four cherubs under her represent music, painting, literature and sculpture, and the four female figures on the base are typical of agriculture, commerce, manufacture and navigation. The whole is of meerschaum, excepting the emblems of the character of each figure, which are of amber. The pipe, though large, would be a fit ornament for a smoking room, the four bowls having each a separate tube.

Some of the other pipes exhibited were really objects of the sculptor's art. Two very large heads representing Mephistopheles and Bacchante were especially beautiful. The stem of each was thirty inches in length, and composed of three hundred pieces of amber. The skulls, faces, heads and figures forming bowls and stems of the various other pipes exhibited by this house were beyond praise as objects of painstaking care and true artistic ability. Many of them were positive curiosities for the skill and attention necessary for their manufacture. In one S-shaped pipe the stem was three-sixteenths of an inch in thickness and six inches in length. To bore this stem was in itself a problem which only patient ingenuity could solve.

The varieties of amber masterpieces and stems were also worthy of attention, especially when it is remembered that the art of making these articles is new to America. The report of the Austrian Commission will fitly serve to give an idea of the estimate in which this exhibit was held by foreigners. They say: The United States of America in the manufacture of pipes and smokers' articles have shown great progress. They were really wonderfully worked objects in this line on exhibition, and although executed by the assistance of foreign workmen, who in this branch of industry have found employment in the United States, nevertheless, it is to be regretted that we must acknowledge and realize the fact that the most important and noteworthy exhibit in this line was that of F. J. Kaldenberg, of New York, whose object, as much by regularity of style as fineness of finish, is to make as superior an article as any of his old established European predecessors in business.

The Kaldenberg presentation pipe for the International Exhibition of 1876. (Courtesy Tinder Box International Ltd.) $10,000-25,000.

With the influence of the clay on meerschaum and as will be seen later, on the briar as well, the twilight of the Nineteenth Century brought the advent of the long, slim, elegant pipe shapes as we know them today. GBD progress in that era aptly sums up what transpired within the entire industry: ". . . the GBD meerschaum still occupied an important place in the GBD collection. Another 100 models in inlaid cases, mainly plain, but including a few very beautiful carved heads. To these we must add a number of intricately carved cigarette holders in meerschaum and amber . . . models as the billiard (néogène) so very popular in the twentieth century were only briefly featured. Bent stemmed pipes, a few with square stems and some with flat or oval stems, were starting to be favoured, also models with what was known as a heel at the base of the bowls were popular . . . the carved meerschaums, in a class of their own, would fetch anything over 75 francs (£3) and 100 francs (£4) was not uncommon for a special model . . . the popular billiard (néogène) was then known as a marseillaise, a ribolboche (or zulu) existed, as did hongroises (hungarians) and haitis (square or round stems), to this day still known as such. Certain curved or bent styles were known as basques and some of them were turned out with a 'crochet' or hook at the base of the bowl. This hook also figured on certain substantial sized straight lay-backs."[51]

It has been said that the cigarette smoking craze in the United States had a significant effect on the meerschaum industry. Further, "meerschaum is no longer so fashionable as it once was. Probably the numerous cheap imitations, the extreme readiness to break of the genuine article, and the fantastic designs into which it was fashioned, have led to its decadence in public estimation."[52] I am not so dogmatic. All good things come to an end at some point in time. The delicate and slight, even the massive, the bulky meerschaums were impractical, never fashioned for a hurried Twentieth Century industrialized world of assembly lines, crowded buses and elevators . . . and middle class income. But the magical and captivative nature of meerschaum has kept its popularity alive and the ranks are now augmented by a new breed of young, curious and enlightened collectors committed to preserve the cult of the White Goddess!

"A perfect Meerschaum pipe is decidedly one of the choicest and rarest gifts of the gods; but like all choice and rare gifts, it is a source of considerable anxiety to the owner. Like women, its, 'name is frailty.' As originally taken in hand, and presented to the lips, nothing can exceed the lovliness of its looks—its delicious smoothness, its graceful rotundity of form, and apparent innocence from everything that can tarnish a reputation. But, alas! you take it as you take a wife, 'for better and for worse;' and again, alas! it does not fare better with the smoker and his Meerschaum than with man and wife."

The Smoker's Guide, Philosopher and Friend
   A Veteran of Smokedom, 1876

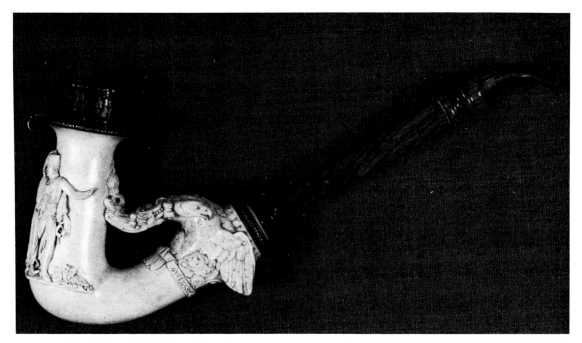

This meerschaum from the third quarter of the 19th century is signed Medetz of Budapest. Stem is turned wood with amber mouthpiece. Turreted lid has Hungarian coat of arms and four silver tokens: a pair of horses, Mercury, a train and a ship. On the shank are inscribed the words: **A Has Zamer.** Purchased in 1978 for $475. (Author's collection) $400-800.

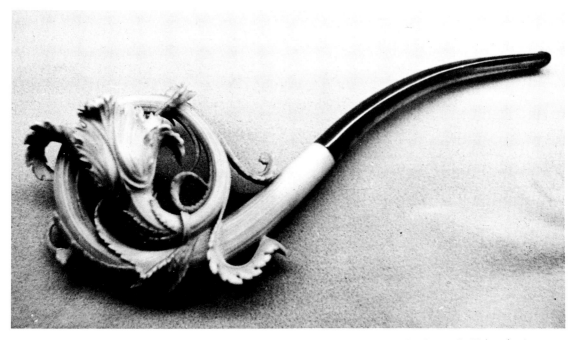

An elegant floral pipe which won a gold medal in Paris, 1889. (Metromedia Collection, Duke University Art Museum, Durham, NC) $500-1000.

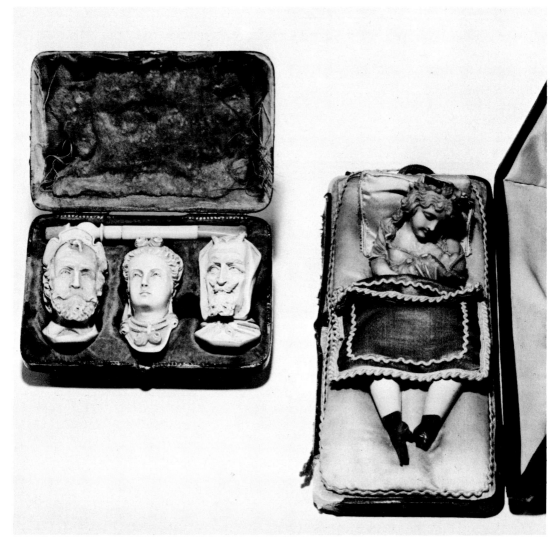

Two cased items from the PB 84 auction. On the left, a set of cigarillo holders representing Faust, Marguerite and Mephistopheles in a fitted leather case. It sold for $400. On the right, a set of cigarillo holders consisting of a nursing mother, reclining under a coverlet, and two miniature holders in the form of legs. This piece sold for $350. (Copyright PB Eighty-Four, New York) Left: $300-600, right: $200-500.

Opposite page:

This gentleman in a very full beard and feathered hat measured out at 8¾" in height and $375 in cost at PB 84. (Martin Friedman Collection) $400-800.

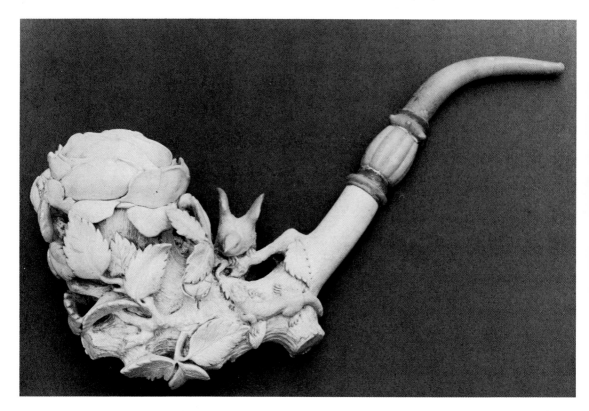

The Rose. Foliate sculpture surround this massive rose as lizard stalks a bird alighting. Pipe measures 15⅞" long, 8" high, including amber ferruled stem. (Herbert G. Ratner Jr. Collection; photograph by Terry DeGlau) $400-700.

Bejewelled sultans in meerschaum from Hungarian craftsmen, 19th century. The three heads are mammoth and have a common bond in style and encrustations. Left and right from the US Tobacco Company Museum; center, the Irving Landerman Collection. $250-450.

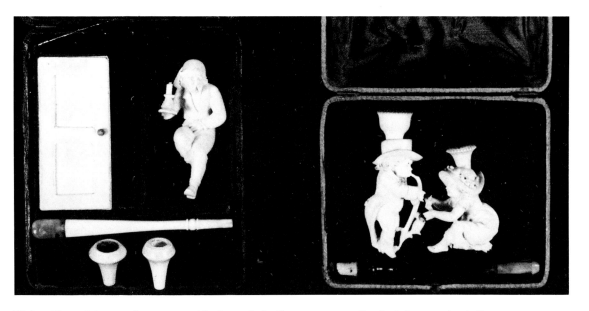

High caliber mini-meerschaum sets evidently made by the same carver. On the left, a cased set of man impatiently awaiting the use of the bathroom. On the right, a youthful couple, boy with pipe and girl with match! (Man and wife from the Irving Landerman Collection: boy and girl (from author's collection) $300-700 each.

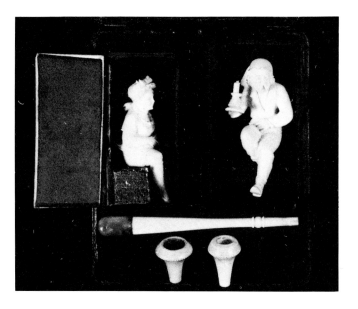

Nubian in Mexican straw hat. (Courtesy Douwe Egberts' Pipe Room, Utrecht, Holland) $300-500.

FOOTNOTES: CHAPTER 4

26. Tobacco Talk, *op. cit.,* p. 28.
27. E. Reid Duncan, "Mysteries of Meerschaum," *Pipe Lovers,* January 1949, p. 8.
28. Ibid., p. 9.
29. *Idem.*
30. Vogel, *op. cit.,* pp. 109–110.
31. Helmuth Aschenbrenner, Tabak von A bis Z (Martin Brinkmann AG, Bremen, 1966) p. 80.
32. Ibid., p. 79.
33. The Great Exhibition—London 1851 (Bounty Books, New York, 1970) p. 17.
34. J. W. Cole, The GBD-St. Claude Story (Cadogan Investments Ltd., London, 1976) p. 11.
35. E. Cardon, Le Musée du Fumeur (Maison E. Cardon et Illat, Paris, 1866) p. 235.
36. Roger Fresco-Corbu, "The Era of the Meerschaum Pipe," *Country Life,* November 10, 1960, p. 1100.
37. Tobacco Whiffs for the Smoking Carriage, *op. cit.,* p. 28.
38. Encyclopédie du Tabac et des Fumeurs, (Éditions du Temps, Paris, 1975) p. 419.
39. Fritz Morris, "The Making of Meerschaums," *Technical World,* April 1908, pp. 194–195.
40. *Long Island Daily Press,* January 24, 1936.
41. Cardon, *op. cit.,* p. 241.
42. Penn, *op. cit.,* p. 167.
43. W. A. Brennan, Tobacco Leaves (George Banta Publishing Company, Menasha, Wisconsin, 1915) p. 158
44. E. R. Billings, Tobacco: Its History, Varieties, Culture, Manufacture and Commerce (American Publishing Company, Hartford, 1875) p. 151.
45. Fume, *op. cit.,* pp. 136–137.
46. Samuel Bevan [Cavendish], To All Who Smoke! A Few Words in Defence of Tobacco; or, a Plea for the Pipe (Baily Brothers, London, 1857) pp. 60–61.
47. Cole, *op cit.,* p. 16.
48. *Idem.*
49. Fume, *op. cit.,* pp. 134–135.
50. Fairholt, *op. cit.,* p. 196.
51. Cole, *op. cit.,* pp. 16–18.
52. J. W. Cundall, Pipes and Tobacco Being a Discourse on Smoking and Smokers (Greening & Co. Ltd., London, 1901) p. 33.

Chapter 5

THE WOODS

"Clay, meerschaum, hookah, what are they
That I should view them with desire?
I'll sing, till all my hair is grey,
Give me a finely seasoned briar."

From "A Ballade of the Best Pipe"
    Robert F. Murray

As with meerschaum, the inauguration of the briar pipe is enmeshed in an assortment of dubious tales but there are some confirmed data about wood pipes which will be revealed in this chapter. All kinds of hardwoods were used for smoking pipes around 1750 with the widespread acceptance and use of tobacco. In one German thesis on pipe manufacturing, 27 exotic woods were recommended for pipe carving, each described with their peculiar properties since again the Germans experimented the most with wood: acacia, alder, ash, birch, boxwood, buckthorn, cedar, cherry, elder, elm, hazel, hornbeam, lava, linden, maple, mulberry, olive, poplar, sycamore and walnut, inter alia.[53]

European literature speaks of an early petition from two pipe-makers from Geislingen near Ulm, to the Council of Ulm. Herr Schaeffler, a weaver and Herr Fassbinder, a cooper, petitioned on May 22, 1695 for permission to go into business but the request was denied.[54] By 1715, there were already 50 craftsmen in Ulm; a second request was submitted and denied. By 1733, the Ulmer or Schwaebisch Maserkopf, that gnarled, speckled, burled, grained, mottled and knotted style wood pipe, a name coined by Jackob Gloecklen of Ulm, had its devotees.[55]

In Ruhla, a certain Simeon or Simon Schenk had begun to make wooden pipe heads in 1739 for what would eventually be styled the Thueringer aufsatzpfeife, a variation of the gesteckpfeife, meaning a head-piece with ornamentation. It is also believed that since there was commerce and traffic between Ulm and Nuernberg, the latter town may have also been making wooden pipe bowls, and at least with certainty, long push stems for the Ulmer. Annals record that wood carvers of Goettingen added a nuance to the Ulmer boxwood pipe bowl in the same decade—a meerschaum lining. At the Frankfurt industrial fairs in 1773 and 1776, the Ruhla artisans Wagner and Hellman displayed their meerschaum pipes; concurrently, Johann Bickeling, Johann Peter Demont, Anton Erdman, Georg Paulus Leinberger and Xaverius Oexlein of Ruhla exhibited an assortment of finished, unfinished, unlined and tin and pinchbeck lined pipe bowls, gesteck—and aufsatz pipes of root woods.[56]

The Ulmer pipe bowl was a specific style, made out of a single piece of wood, usually bearing a helmet like or dome-shaped lid reminiscent of Sixteenth Century footsoldiers, a lid with ornamented cutwork. To the eye, top of bowl and neck are of

Woods of Europe. On the left, from the Tinder Box International Pipe Collection, a root pipe in natural form measuring 34" and a cryptic note in the bowl with the following: *Carl Holm, Jylland, 19 April 1914;* in the center, a 22" Black Forest pipe of root wood with metal fitments; right, a handsome carved relief pipe with light wood floral appliqué ornamentation, dog and bird, measuring 27". (Pipes in center and at right are form the Pozito Collection, Madrid) $200-600.

A Dutch forester's wood turned pipe from the 1790's. (Courtesy Douwe Egberts' Pipe Room, Utrecht, Holland) $100-200.

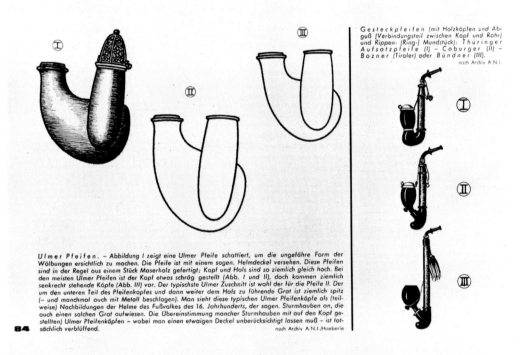

On the left, the Ulmer. I is a finished product. II is the standard Ulmer with forward leaning bowl and III is a refinement with a perpendicular bowl. To the right, gesteckpfeifen varia. The three shown are wood bowls and reservoirs, but the appellations are the same for porcelain and meerschaum. I is the Thueringer aufsatzpfeife; II is a Coburger, after the city of Coburg; III is the Bozner (Tiroler) or Buendner style. (Reproduced from H. Aschenbrenner, TABAK von A bis Z, Bremen, 1966)

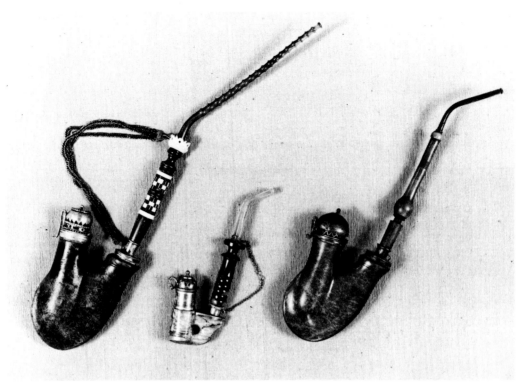

Two 18th century Ulmers, in boxwood, flanking a miniature 19th century Russian interpretation in olive wood with metal fitments, horn stem and amber mouthpiece. (Author's collection) $150-500.

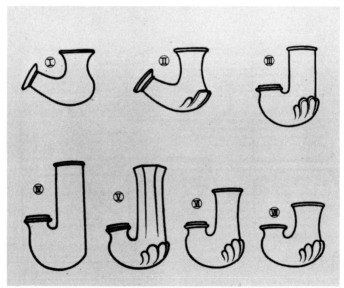

Sketches of the Hungarian family. I and II are Kalmasches. III and IV are Debrecens, III more ornate than IV. Styles V, VI, and VII are refinements identified as Ragoczy. Note the similarities as well as the slight but distinguishing differences for classification. Were V not so tapered in the middle, it would have been classed as a panelled Debrecen. (Reproduced from H. Aschenbrenner, TABAK von A bis Z, Bremen, 1966)

An 18th century Kalmasch in walnut burl—masuline and massive! (Courtesy H. F. and Ph. F. Reemtsma, Hamburg) $200-500.

the same elevation . . . bowl slightly inclined forward; another prototype included a perpendicular bowl. The shank was fitted with long stems of antler, horn, cherry and ebony, many with mother-of-pearl, bone and ivory marquetry inlay. The wood most used for the Ulmer was boxwood, a fine textured light yellow hardwood, found in Europe and Asia, a wood with no grain limitations, excellent for carving, generally available in only small logs, quite costly and sometimes giving off a sweet odor. The Ulmer style was so popular that it was also fashioned in meerschaum.

Akin to the Ulmer was the Hungarian. German exactitude has, once again, resulted in three precise classifications generally agreed to by pipe ethnohistorians:

    1. Kalmasch, a derivative of the Turkish chibouque, shaped like a kettle, cauldron or inverted bell.

    2. Debrecen, named for the town in East Hungary famous for its red clays; another descendant of the chibouque; a cylinder, chimney-pot, plain or panelled.

    3. Ragoczi or Ragoczy, imitative of the Debrecen with a slight outward taper or widening at the top of the bowl.

To insure a more complete understanding of all pipe styles, I have included a facsimile drawing of gesteckpfeifen. Remember that the gesteckpfeife is a style, a class, a genre; the word should not conjure up thoughts of specific materials. As mentioned in Chapter 3, this design was used for porcelain bowls. It was very much used for meerschaum and wood bowls; and, both meerschaum and wood reservoirs were not uncommon gesteckpfeife components. No glossary of style can

Left: A boxwood and horn pipe from a St. Claude manufactory in the early 1800's. It has the flavor of art Deco or art Nouveau. (Courtesy le Musée de la Pipe, St. Claude) $150-300. *Right:* An early 1800 boxwood and horn pipe from St. Claude. This one is of the gesteckpfeife genre with wood abguss. (Courtesy le Musée de la Pipe, St. Claude) $100-300.

authenticate the effects of the passage of time and the free expression of individual artisans unfettered by rules of design and patent law. Many of my own antique wood pipes do not conform to any of the aforementioned styles, but with good reason. Between 1797 and 1812, in Ulm alone, there were a record 45 pipe-makers, making Ulmer bowls. More significantly, from 1739 to about 1840, 114 different permutations of the Ulmer have been identified.[57]

One elegant description of these hardwood pipes appeared in a mid-Nineteenth Century Scottish journal: "The Germans have perhaps experimented more profoundly in pipes than any other people. They long used a beautiful pipe carved by the herdsmen and peasants of the Black Forest from the close-grained and gnarled root of the dwarf-oak. The wood is hard enough to resist the action of the fire, becoming but slightly charred by years of use. The carvings represented sylvan

A Debrecen refinement from Vienna, hallmarked 1807-1810. This pipe has surmounted silver ornamentation of lion rampant, crown and retaining chain. Attributed to Franz Josef I, Emperor of Austria and King of Hungary. (Ed Clift Collection) $200-500.

Four extremely fine examples of bas-relief 19th century wood bowls with silver ornamentation and turned wood and horn stems. (The late Hermann Thiede Collection) $200-500.

A crudely sculptured relief wood pipe bowl depicting revolutionary soldiers in the leisure activities of card-playing and drinking. It has, however, an ornate silver shank and silver domed pierced lid. (Rothschild Collection, courtesy le Musée de Grasse) $200-500.

This wood bowl is reminiscent of the fanciful von Schwind sketches for the meerschaum! Here, four horsemen cross a drawbridge while a man, in bas-relief, fishes near the castle. The pipe lid is the castle's tower. The bowl is lined in tin. Inscription on the bottom reads: **Freigt '88.** This pipe was purchased in 1967 for $175. (Charles P. Naumoff Collection) $300-700.

Exceptionally detailed and ornate wood pipe from Germany with horn and reh stemwork and ivory articulation from the 19th century. (Courtesy Austria Tabakwerke, Vienna) $300-1000.

Eighteenth-nineteenth century European woods. Top left, cane handle pipe; top right, a Kalmasch mottled boxwood with silver figural ornamentation of the hunt; in the center, a massive bowl of French briar; lower left, a stylized panel with wooden lid; lower right, a lava wood bowl with bas-relief of serpentine arches. (Center pipe from the Irving Landerman Collection; all others from the US Tobacco Company Museum) All, $200-600.

Two finely carved specimens of the wood turner. To the left, a turbaned Easterner with two naked men surmounted, one tugging at the mustache; to the right, a helmeted man looking skyward above a smiling lion cub. Both are 19th century French. (Rothschild Collection, courtesy le Musée de Grasse) Left: $300-500, right: $200-400.

scenes—boar-hunts, recontres with wolves, sleigh-driving, fowling and the exploits of robbers. Not unfrequently the subject was an illustration of ancient German literature, as a scene from the story of Reynard the Fox."[58]

Soon thereafter came the début of erica arborea, the heath tree, briar. The derivation of the word briar is a corruption of the Latin brugaria, the Celtic brug, the French bruyère. Further vulgarization is found in these Nineteenth Century appellations: Baumheide in German; Boomheide or Groote-Heide in Flemish; Scopa-arborea, Scoponi-da-bosca and Stipa, in Italian. Characteristically, this material has porosity, hardness, weightlessness, a fine grain, resistance to heat and is most ideally suited for a smoking pipe.

As one legend of its accidental discovery goes, a French pipe-maker on a pilgrimage to the birthplace of Napoleon Bonaparte at Ajaccio, Corsica, accidentally dropped and shattered his meerschaum pipe. In search of smoke, he commissioned a Corsican wood carver to make a pipe. The carver used a briar burl found in ample supply in all countries bordering the Mediterranean. The Frenchman brought back samples to Saint-Claude in the Jura Mountains and eureka! . . . the birth of the briar.

There is much symbology in this sculptured wood bowl with wood abguss and drain plug. It depicts the retreat of Napoleon Bonaparte from Russia. Throughout, there are the accoutrements and implements of war, but special attention should be given the lid which is truly prestigious. The bowl is lined in clay. Purchased in 1965 for $175. (Charles P. Naumoff Collection) $300-900.

This 18th century pipe from Afghanistan is multi-media personified. It is the blending of East and West. The wood bowl takes the form of a Turkish slipper, lined in meerschaum. The lid is surmounted with the head of an Oriental whose neck oscillates with every move of the pipe. The exterior of the bowl is encrusted with turquoise, rubies and inlaid copper wire. It has a long and a short stem, both encircled with coral and ivory, a mouthpiece of amber. The overall length is approximately 50". (Courtesy le Musée du SEITA) $300-700.

This rather whimsical sculptured pipe is either a lion or a bear on a sled. It is a 19th century expression and has silver fitments, retaining chains and an enamelled blazon of lion rampant and the letters, *P.F.-V.R.* (The Rothschild Collection, courtesy le Musée de Grasse) $250-600.

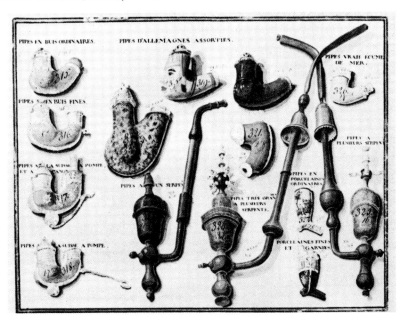

An extract from an 1807 catalogue of the St. Claude firm, Jeantet. In 1806, one year earlier, at the general exposition of that village, of the 184 sample items of wood on display, only four were pipes. This facsimile page displays wood, porcelain and meerschaum.

In a recent American textbook on tobacco and pipes, it was a French envoy, travelling from Geneva who lodged one evening in Saint-Claude. Upon discovering that his meerschaum pipe was shattered, he sought to repair it. One of the village wood turners crafted a replacement pipe out of briar burl, a wood sample obtained from the Mediterranean coast.[59]

There are yet other interpretations. In the village of Saint-Claude, capital of Haut-Jura, there is the Abbey of Condat which has attracted pilgrimages since the end of the Eighteenth Century. There, the local artisans had been turning woodware, making religious articles, necklaces and other relics for the tourist. They made occasional pipes and the principal woods with which they worked were boxwood, cherry, pear and arbour or arborvitae; they were not heat resistant, they released a disagreeable odor which pipe smokers found offensive. Around 1854, a dealer in wood from Midi had the idea to test briar and with time and experimentation—success.[60] Jules Ligier fixed the beginning of the Saint-Claude briar industry epoch at 1857.[61] J. W. Cole, referring to Messieurs Ganneval, Bondier and Donninger says: "Clear-sighted as well, they realised the great possibilities of a new material, racine de bruyère (erica arborea), which was already being utilised with success in St. Claude in the Jura, and by 1855 Briar GBD pipes were sold alongside the GBD meerschaum."[62]

There are no pipe-smoking Delphic oracles to resolve the distinguishable yet insignificant differences among the many and varied scenarios, but suffice to say briar was a phenomenon of the last half of the Nineteenth Century. One thing is certain—the design for such a small smoking apparatus as we now know the briar was not devised in Ruhla, Ulm, Coburg or Nuernberg; it is an adaptation, a direct descendant of the ancestral clay form. Credit must be given to the French and especially Saint-Claude for the briar's resemblance or similitude to the clay and their early manufacturing successes with it.

A carved briar head of a cavalier, late 19th century. Only an adept carver could have been able to skillfully evoke such an aged and wearisome countenance. (Charles P. Naumoff Collection) $200-350.

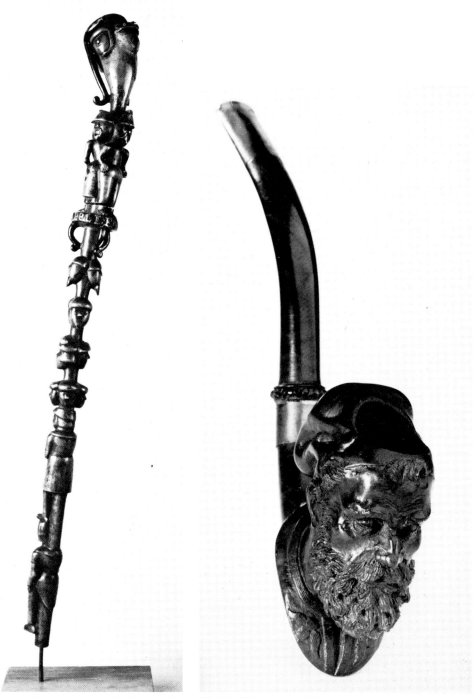

Left: This peculiar pipe of wood is from the last half of the 19th century. Throughout the stem and on the bowl are Oriental personages wearing clothes of the Western World. It is believed that this pipe, called *Pipe de Compagnonage* (a pipe of comradeship or fellowship) may have been carved in Indochina for a Frenchman. It bears the inscription **Chapell André 23**. (Courtesy le Musée du SEITA) $400-700.

Right: Carved briar of St. Claude from the last quarter of the 19th century. (Courtesy Douwe Egberts' Pipe Room, Utrecht, Holland) $200-400.

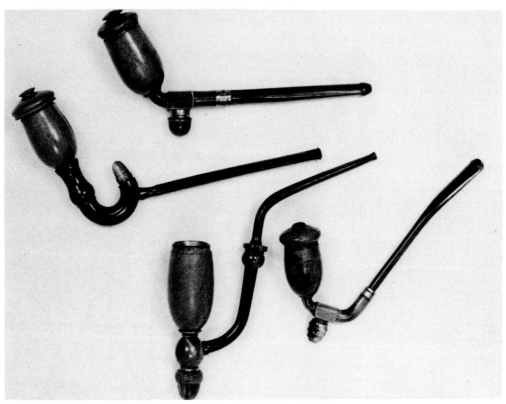

A group of St. Claude boxwood and horn pipes from the first half of the 19th century. Disregarding the drain plugs, there is evidence of the evolutionary trend toward the classic standard shape briars of the latter half of the 19th century. (Courtesy le Musée de la Pipe, St. Claude) $100-250.

In this same era, American enterprises arose to the occasion and entered the briar pipe industry: S. M. Frank & Company, Reese Brothers and the Linkman Company, both of Chicago, Kaufman Brothers and Bondy. William Demuth & Company tooled up to make briar. In England, the first to manufacture briars was Joseph Koppenhagen of London in 1862. Frederich Kapp, Soho, an early manufacturer of meerschaums, took notice of a popular pipe made of briar root and "opened a branch house at 53, Grafton Street, Dublin, at the windows of which the natives are much astonished and interested at seeing the above manufactures conducted, this being the first and only establishment of the kind in Ireland."[63]

Returning to France, in 1841 Saint-Claude accounted for only three factories dedicated to woodworking with fewer than 20 workers in each. In 1892, the number of plants devoted to pipe manufacture rose to 66 with two foremen and 1600 workers, not counting 500 women and 200 children.[64] By then, bone, amber and ivory stems were dethroned, horn survived and with electricity and steam engines, the advent of vulcanite. For an idea of selectivity, I turn again, to Cole. "In briar they presented a basic range of some 160 models in top quality and the same number in what was then called 'mixte.' Then came about 60 models with 'grands ambres' also in each grading. Following these were some 50 shapes in each quality of brazil horn screw fitted with oval bore, with a cheaper range horn fitted (again 100 shapes in each grade). Continuing further, we find a range of 100 models in each of the two qualities fitted with ivory, and also similar models with rubber stems . . . a wonder collection comprising about 1300 models in briar alone . . ."[65]

A sculptured briar from St. Claude at the turn of the 20th century. (Courtesy le Musée de la Pipe, St. Claude) $100-200.

An assortment of 19th century French comical character wood pipe bowls. (Courtesy le Musée d'Intérêt National du Tabac, Bergerac) $75-200.

Assorted wood pipes have appeared as novelty bottles; as a caricature pipes—sculptured heads with grotesque supersize proboscises mounted in self-supporting stands, souvenirs of the French Barbizon era; and, commemorative pipes of the Boer Wars and our own Civil War.

The rest of the briar story is universally known. Perhaps one Twentieth Century American anecdote is worthy of mention. Noted American sculptors have been more than peripheral to the pipe. Mr. Robert L. Marx of Mastercraft Pipes, New York City, made a significant contribution to the collector world some thirty years ago with the Marxman Heirloom Collection, now on display at the United States

The fanciful French pipe bowl on the left is the famous sabot, the wooden shoe, with the added attractions of a high relief bird and surmounted head of what appears to be a hog or pig. On the right, an even more whimsical combination of cannon and rooster. Both date back to the era of the French Revolution. (Courtesy le Musée du SEITA) $75-150.

Popular art of the late 19th century in France. Grotesque carvings of amateur whittlers. (Courtesy le Musée du SEITA) $50-100.

This pipe from the Heide Collection has a multinational flavor and bears the following attribution: Cuban, carved from a tree branch, Chinese workmanship with European influence. The ornamentation is in brass. (Charles P. Naumoff Collection) $100-200.

An early 20th century briar...a setter as graceful and delicate as any from the turn of the century. (Courtesy Museum Chacom, St. Claude) $200-400.

A sleek briar with amber mouthpiece, early 20th century. (Courtesy Museum Chacom, St. Claude) $100-200.

A graceful and unusual early 20th century briar from France. (Courtesy Museum Chacom, St. Claude) $200-400.

An early 20th century figural briar. The cow's eyes light up when the bowl heats. (Courtesy Museum Chacom, St. Claude) $150-300.

Tobacco Company Museum, Greenwich, Connecticut. During the 1940's he blended his business acumen, his love for art and an appreciation of briar as a medium for the artist by persuading renowned artists to sculpt briars, some for resale, some for posterity. As a result, there are today mammoth, full and bas relief carved briars as Old King Cole, Bacchus, the Huntsman, Fox and Pheasant, the Forty-Niner, Stalin and Roosevelt at chess. The carvers who collaborated were Jo Davidson, Leon Cutler, Louis Ted Shima, Charles Kopp, R. D. Watts, Edwin F. Drake, G. A. Griffin, Hetzer Hartsock, Garlow and Cecil Howard, a former president of the National Sculpture Society and Pipe Smoker of the Year in 1947 for his briar masterpiece of a reclining seminude.[66] The Marxman Heirloom Collection eventually totalled 80 carved briars.

Woods have come a long way! It is said that there were some carved wood pipes in the collection of His Late Royal Highness the Duke of Sussex, K.G. "but however well they may have been carved, they were no use for smoking because the right

This bird's nest sculpture with horn stem is a St. Claude briar from 1920-1930. (Courtesy le Musée de la Pipe, St. Claude) $100-250.

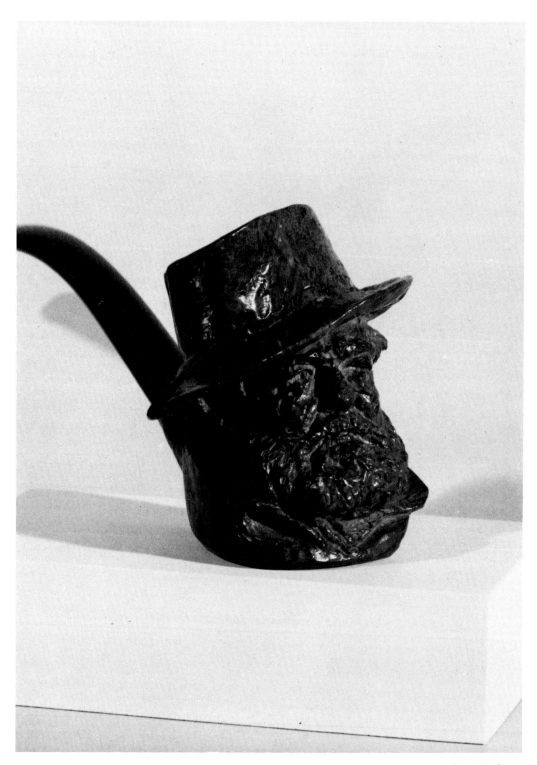

A carved briar from the Marxman Heirloom Collection...self-portrait of the artist Jo Davidson of New York. This pipe was carved for Robert L. Marx in the 1940s. (Courtesy US Tobacco Company Museum) $150-300.

Two mid-20th century carved briars sculpted and autographed by Garlow. Above, this 11" pipe of two Indians approaching a dead buffalo was purchased for $425 in 1978. Below, the 10¼" pipe of a Civil War soldier firing his pistol from defilade is dated 2-52 and was purchased in 1978 for $300. (Copyright PB Eighty-Four, New York) $600-900 each.

A 15" Garlow briar executed in April 1951. Three frontiersmen around a campfire. This pipe was purchased for $600 in 1978. (Copyright PB Eighty-Four, New York) $600-900.

wood for the business was still unknown."⁶⁷ While that was true for smokers, the hundred years prior to the Duke's 1843 auction represented a century, no, the ultimate era of the European wood turner and pipe carver. Adroit, dexterous, proficient and inspired, he transformed one of nature's raw materials into art treasures of lasting duration. And that was soon forgotten for it was briar which then took command. In 1885, *Notes and Queries*, a British periodical, stated: "Thirty years have enabled these intruders to destroy short clays, ruin meerschaums, and even do much mischief to the venerable churchwarden."⁶⁸

During the period 1907-1910, Joseph Schneider, descendant of a long line of Viennese wood carvers, executed this masterful presentation pipe for exhibition at the International Hunting Exposition in Vienna, 1910. The pipe is 70 cm tall, made of Swedish birch and accounts for several hundred figures. The pipe lid is a Hapsburg motif—Rudolph of Hapsburg is loaning his horse to a priest attempting to cross a raging river to give last rites to a dying man. At the lid's base are the words: **Hail to the Pious Hunter** and the coats of arms of lands formerly under the Crown. The bowl depicts Rudolph as the sovereign and Field Marshal who, following the battle of Duernkurt, stands before the vanquished and dead Otto Kar and his soldiers, giving a speech on earthly power. The Emperor is surrounded by warriors, horsemen at closed ranks and prisoners being led away. On the reservoir is inscribed Rudolph's speech, birds in flight to indicate transience and a teetering ship as allegory on the uncertainty of life. The inscription follows: **What is our life here below? It is a battle without rest and peace. Death is bestowed on us all while we strive for eternity.** In 1910, Schneider was offered 10,000 kroner, later as much as 80,000 kroner for the pipe but he rejected all offers! Schneider died in the spring of 1941 and his son Hermann has seen fit to make this pipe public. (Courtesy Austria Tabakwerke, Vienna) $3000-7000.

FOOTNOTES: CHAPTER 5

53. G. M. Raufer, Die Meerschaum und Bernsteinwaren Fabrikation (A. Hartleben Verlag, Wien, 1876) pp. 117–128.
54. Aschenbrenner, *op cit.*, p. 82.
55. Adolf Haeberle, Die Beruehmten Ulmer Maserpfeifenkoepfe (Verlag Otto Wirth, Amberg/Oberpfalz, 1950) p. 22.
56. Aschenbrenner, *op. cit.*, p. 82.
57. Haeberle, *op. cit.*, pp. 24–25.
58. Chambers Edinburgh Journal, February 9, 1856.
59. Carl Ehwa Jr., The Book of Pipes and Tobacco (Random House, New York, 1974) pp. 104–110.
60. L. Vincent-Coutier, Saint-Claude et l'Industrie de la Pipe (Imprimerie Moderne, Saint-Claude, 1921) p. 4.
61. La Pipe Bruyère, Saint-Claude 1856–1956 (Imprimerie Moderne du Courrier, Saint-Claude) p. 18.
62. Cole, *op cit.*, pp. 1–2.
63. Tobacco Whiffs for the Smoking Carriage, *op. cit.*, p. 28.
64. La Pipe Bruyère, *op. cit.*, p. 22.
65. Cole, *op. cit.*, p. 16.
66. John Giaccio, "First Take an Old Hacksaw . . ." *Wonderful World of Pipes*, Vol. 1, No. 2, 1971, pp. 18–19.
67. Compton Mackenzie, Sublime Tobacco (Macmillan Company, New York, 1958) p. 253.
68. *Ibid.*, p. 254.

Chapter 6

ETHNOGRAPHICA I: NEAR AND FAR EAST

"My nargileh once inflamed,
Quick appears a Turk with turban,
Girt with guards in palace urban,
Or in house by summer sea
Slave-girls dancing languidly,
Bow-strings, sack, and bastinado,
Black boats darting in the shadow;
Let things happen as they please,
Whether well or ill at ease,
Fate alone is blessed or blamed."

From "The Smoke Traveller"
Irving Browne

"Of course, everybody has seen, in shop windows at least, the mighty Hookah or Hubble-Bubble, a sort of chemical laboratory invented by some ingenious Oriental, begotten in melancholy, reared in idleness, and naturally condemned to everlasting fatuity."[69] To the collector, the pipes of the Near East are not to be so derided since the land of Smokers evolved two unique pipe genres, both handsome implements of the smoking culture.

The first is the Turkish peasant or commoner's chibouque (Tchibouk, Tschibuk, Chibook, Tchibukdi) described as a clumsy, barbarously splendid affair, consisting of a separable bowl (khagar) of reddish brown clay (shown in Chapter 2), a long wood stem and a small ovoid mouthpiece. The clay, sometimes identified as Samianware, is baked and like old Roman pottery is of a terracotta hue. The mouthpiece is oval for "a Turk usually takes a better hold of the mouthpiece with his lips, and inhales more strongly than an English smoker."[70] Two descriptions suffice to describe its romance: "The cherry stocks come from Persia by Trebisond; they are brought to Constantinople in pieces of about 2 feet long; & after being set straight, are dressed & polished with infinite care. They are united into sticks generally 5 or 6 feet, though some are as long as 12, & the junction is so skillfully concealed with the bark, that in a well made pipe it is impossible to discover it. When repolished they are ready for sale, being left unbored until the merchant finds a purchaser. From 30, to 100, piastres is the usual price, but it differs according to the length, size, & fineness of the bark, & dark ones are preferred."[71] The almost imperceptible manner by which the stems of not only cherry but also jasmine were joined had to have been the result of skill and ingenuity and no doubt commanded a handsome price. Professor Arminius Vámbéry commented on the very meticulous Turkish: "The pipe-stems, the bowls, and the mouthpieces . . . are chosen with the greatest care and kept in order with the most scrupulous attention. The bowl must come from the manufactory of Hassan, in

Ladies' chibougue, heavily enamelled and filigreed. (Metromedia Collection, Duke University Art Museum, Durham, NC) $150-300.

the faubourg Findekli. The long stem of jasmine wood, with its satin-like bark, must have grown amid the shrubberies of Broussa. The mouthpiece must be made of the clearest and most transparent amber, cut in the most fashionable shape; and the tube on which it is screwed must have been turned by the most cunning workmen."[72]

As the reader can surmise, this so-called peasant pipe was not relegated just to the common man. The wealthy smoked the chibouque, but theirs were distinguished by silk and gold ornamentation and precious stones. The bowl surface was either smooth and polished or etched with scrolls and foliates. Domed ornate pierced covers were not uncommon, made of fine metal and bejeweled with agate, coral and turquoise. The average length of the chibouque seems to be between four and five feet. Those with silk or embroidered fabric stems were to be moistened to cool the pipe.

The most costly part of the pipe was always the mouthpiece. It was transparent amber according to Vámbéry. Fairholt reported that "the mouth-piece (foom or turkeebeh) is composed of two pieces or more of opaque, light-coloured amber, interjoined by ornaments of enamelled gold, agate, jasper, carnelian or some other precious substance."[73] Forrester's observations on amber at a Turkish bazaar should not be excluded: ". . . there are two sorts: the white, creamy, or lemon coloured amber is most valuable; a large mouth-piece of the very finest is worth from 5,000 to 6,000 piastres, about £50, or £60, the second or yellow kind being more common is comparatively little esteemed, for the perfection of the article consists in its being free from flaws, or spots; & if the tube can be seen through the amber it is considered very inferior. There is a 3rd. sort, valueless from its transparency. It is either real or factitious, often consisting of the scrapings & refuse melted into lumps & manufactured into cheap mouth-pieces. This portion of the Turkish pipe is frequently adorned with precious stones, enamelling, or carved wood, the cost of those generally exposed varies from 20, to 1500, piastres . . ."[74] The Egyptian and Algerian variety of the chibouque, I might add, is almost indistinguishable from the Turkish. And, these countries have used a shortened version of the chibouque with a metal bowl called copoq.

The chibouque was so much a part of Turkish diet that the smoking customs of that land have been the subject of ribald humor. "The Turks have sunk into luxurious even if inglorious inactivity and ease, preferring the loss of provinces to the loss of time for smoking."[75] Worse, a Persian, Nadir Shah said: "You need not have any fear or anxiety respecting this nation; for the prophet has given them but two hands; one of which is absolutely requisite to keep on their caps, & the other to hold up their trowsers; if they had a third, it would be employed to hold their pipes; they have therefore not a spare for a sword or shield."[76] I believe I have made my point!

A nargileh of chased brass with flexible fringed hose. (Courtesy Austria Tabakwerke, Vienna) $300-500.

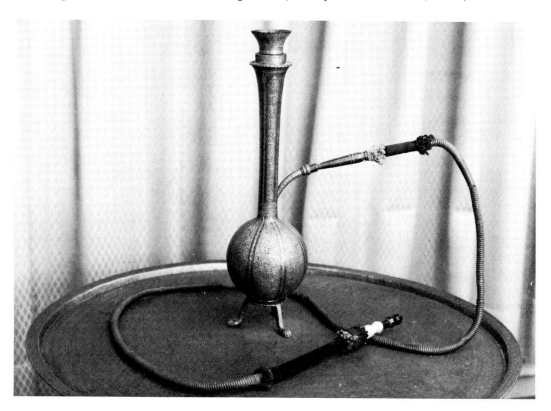

An Iranian silver chased nargileh with turquoise encrustations. The bowl is lined with clay. It rests on a wooden smoking table with ivory marquetry. (Courtesy Austria Tabakwerke, Vienna) $1000-3000.

Near East water pipes have been likened to the nectar and ambrosia of Olympus, ice to champagne, the acme of tobacco smoking for the smoker was prone to add sandalwood, aloe, rose leaves and even hashish or opium to the tobacco in a water pipe. His water pipe is not a mere water pipe—it is of many names, many interpretations, many styles! There are hookahs, nargilehs and kallians and they are as different as the words themselves. The Indian nargileh, nargeeleh or narghile, philologically, stems from the Persian word nargil, coconut—a fruit originally used as a water flask whose shape was copied as a water pipe and continued to be a handheld contrivance. The simplest form entailed three fundamental parts: a cup for the tobacco, a tubular stick and a pot for water. There was a domestic style for home use and a type equipped for outdoor use and travel. A complete nargileh made of

A nargileh consisting of brass chased bowl, lacquered wood stem and brass serkallian lined with clay. (Courtesy Austria Tabakwerke, Vienna) $500-1500.

only one metal is very rare, since the diverse pieces or components are normally made of different materials. Pots have been found in glass, clay, wood and stone, porcelain, gourd and certainly coconut, each with its own peculiar ornamentation. The glass was generally painted; the clay was generally moulded and colored; those of wood or stone were sculptured; and, those of metal were engraved, chased or decorated. Many are feasts for the eye! "The metal fittings of these pipes are frequently of gold and silver; the flexible tubes, from 5 feet to 10 feet long, through which the smoke is drawn, are covered with velvet and encrusted with precious stones and gold filigree work. The waterbottle is of the finest cut glass, and is handsomely decorated with diamonds and other gems."[77] A tripod can accompany the nargileh to keep it erect and free-standing.

Persian kallian in cranberry glass with gold leaf etching. (Courtesy Wills Collection of Tobacco Antiquities, Bristol, England) $300-400.

A serkallian from the period of Qadjar, end of the 18th century, Iran. The upper cylinder is polychromatic enamelled copper with four insets, two of women, two of flowers. The socket is of wood, embossed in foliated scrolls. [Note: the bowl is inverted in this picture.] (Courtesy le Musée du SEITA) $300-500.

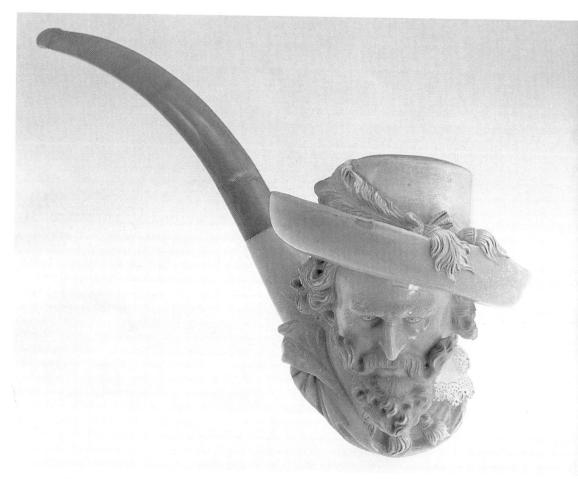

The 19th century meerschaum pipe is a high relief bust of a hirsute gentleman in late 17th century garb. He wears a soft hat, adorned with ostrich feather and tassels and a smock with lace turn-down collar. (Courtesy Douwe-Egberts' Pipe Room, Utrecht, Holland) $350-600.

Opposite page:
Porcelain pipes and bowls from the Jean Charles Rhein Collection. $200-700.

1. Meissen, around 1820.
2. German, around 1850.
3. Costume from Canton of Thurgau, Switzerland, around 1850.
4. Costume from Canton of Schwyz, Switzerland, around 1850.
5. Costume from Canton of Berne, Switzerland, around 1850.
6. Silhouette from Zurich, Switzerland, 1852.
7. Armorial from Germany, 1838.
8. On Front, Tuebingen, 1819; reverse: 69 names of associates.
9. German, around 1850.
10. German, around 1800.
11. German, Nineteenth Century.
12. Biedermeier style, reservoir painted en suite, around 1850.
13. Another en suite in Meissen, around 1850.
14. Berlin, Free Mason pipe, around
15. German, around 1850.
(Courtesy Tabac Rhein)

1

2

3

4

5

6

7

8

9

10

11

12

13

14

15

An early 20th century example of a simple tube pipe of bamboo from Annam, now Vietnam. (Courtesy le Musée du SEITA) $100-200.

A shisheh, sheeshé or sheesheh is a nargileh of Turkish origin which has a glass vase. The gozeh, a member of this family, has a short cane tube rather than a long flexible hose and it was most often used by the very lowest class of smoker for both tobacco and hemp. The Arab hookah or huqqah, meaning box or basket in Persian, on the other hand, is a free-standing device for divan or floor. It is said that the hookah, a rather ponderous invention, was originally invented in Persia and the concept spread throughout the Near East, followed by the Indian interpretation of the nargileh. One thing is certain, as the Turk was never far from his chibouque, so the worshipper of Allah swore by his beard, his horse, his scimitar and his hookah. All Near East water pipes with flat bottoms then, are hookahs and the Persian knows it as kallian, kalian or qalyan, a water vessel of glass, clay or coconut. The hookah or kallian (small hookah) consists of a vessel for water; a hollow tube made of wood, ebony or silver, adorned with small chains; mounted to that is the pipe bowl, a serkallian, chilloom or chillum of clay or wood. The inner side of the serkallian is generally coated with lime, the exterior can be plain or ornately decorated with enamel or silver. A rigid stem is inserted into the kallian for home use; a marpitsch, a long leather flexible hose, is used when the owner chooses to smoke while travelling. Such tubes have measured from five to ten yards. Legend tells that the Sultan Mahmud II, 1808–1839, gave a certain Pasha Ferik a hookah with so lengthy a stem that ten slave girls acted as bearers and were, therefore, considered a part of that gift!

The kallian and hookah were no less decorative than its relative, the nargileh. "The hookah smoked by the Shah of Persia on state occasions is so studded with

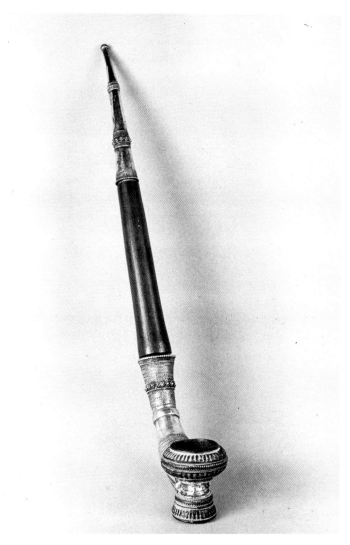

This particular style of pipe has been attributed to Burma, Laos, Cambodia and other locations of the Far East. The silver chased bowl is lined with black clay; the wood stem is finished in silver. (Courtesy le Musée du SEITA) $100-250.

diamonds, rubies, emeralds, etc., that it is worth £80,000."[78] Some kallians required more lavish outlays for their costly composition than one had to pay for a meerschaum of equal caliber. Fume said: "The receiver, a fountain, in the most splendid of these Oriental smoking machines, is usually of cut glass, silver or enamelled gold, and of a size sufficient to contain about three pints of water. In smoking it is about half filled, and a tube, proceeding from the receptacle for the tobacco, reaches nearly to the bottom. The end of the smoking tube also enters the receiver, but does not touch the water. Both these tubes are bound together, and fitted closely to the neck of the receiver, so that no air can enter . . . the person who contrived the gasometer must have borrowed a hint from the principle of the hookah."[79]

To inhale the smoke of a hookah requires strong lungs and the typical gurgling of water caused by the ascending smoke has prompted other names for the hookah: the Anglo-Indian slang hubble-bubble, gurgarri and gurguru! The novelist Balzac

An exquisite 19th century expression in meerschaum with cherubs, garlands and the Austrian double eagle. This pipe measures 12" long and is from the adept hand of Georg Kopp, Dresden. Purchased in 1978 for $1,150. (Author's collection). $1200-2000.

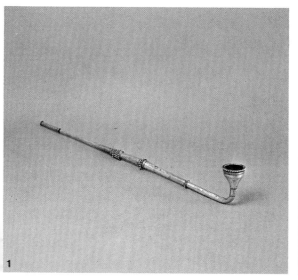
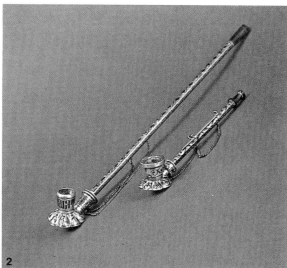

Silver pipes from the Jean-Charles Rhein Collection:
1. Nineteenth century from Cambodia, 50 cm
2. Nineteenth century from Russia, 28 cm. and 14 cm
3. Decorative cigar holder, 12.5 cm
4. Nineteenth century from Iran with gold filigree, 12.5 cm
(Courtesy Tabac Rhein) Top left: $100-300, top right: $100-300, bottom left: $100-200, bottom right: $200-500.

Opposite page:
This meerschaum was made in 1871 and entitled The Marriage of HRH the Princess Louise and the Marquis of Lorne in St. George's Chapel, Windsor. The theme was modelled after an engraving which appeared in a special supplement of the *Illustrated London News,* April 1, 1871. The figures, which include Queen Victoria, are grouped around a pulpit, topped by a silver gilt cap, a coronet, perched on a tasselled cushion. The bowl, 18" long, is shaped like a boat with a blown rose carved at the base. The stem is of two-tone amber with silver gilt bands. This £2,200 ($4,400) pipe is on display at Dunhill's London showroom. (Courtesy Alfred Dunhill Ltd.) $3000-6000.

also waxed prolific on the hubble-bubble by describing it as "a very elegant instrument; its shape is extraordinary and lends a sort of aristocratic superiority to him who uses it under the eyes of an astonished plebian. It is a reservoir bulging in shape like a Japanese pot, which supports a sort of tripod of terra cotta in which the tobacco is burned. The smoke passes through hollow tubes several yards in length, adorned with silk and silver thread, whose end plunges into the vase above the perfumed water which it contains, and in which disappears the tube which comes from the upper chimney."[80] As a final word on the opulence of Near East pipes, a few words uttered nearly one hundred years ago are worth a thousand pictures: "A collection of pipes worth £6,000 ($30,000.00) is no unusual thing with high official and rich private persons in Constantinople."[81]

There are many other colorful smoking implements of republics and principalities of Asia I have intentionally bypassed in order to address the other major influence on the pipe collecting world—the Orient. Writers of tobacco culture have identified various dates for the introduction of tobacco to the Orient. It is adequate, for the purpose of this discussion, to say that tobacco was introduced to China during the Wan-li period (1573–1620) of the Ming Dynasty, although China has claimed the Yuen Dynasty, circa 1300; Japan was to have used tobacco in the Fifteenth Century, smoking was quite common in 1595 and in 1596, tobacco seed was imported and cultivated in Satsuma Province; Korea experienced smoking

Bamboo pipe and tobacco holder, Borneo. Note the trade beads and the U.S. trade dollar dated 1877. (Ken Erickson Collection) $100-200.

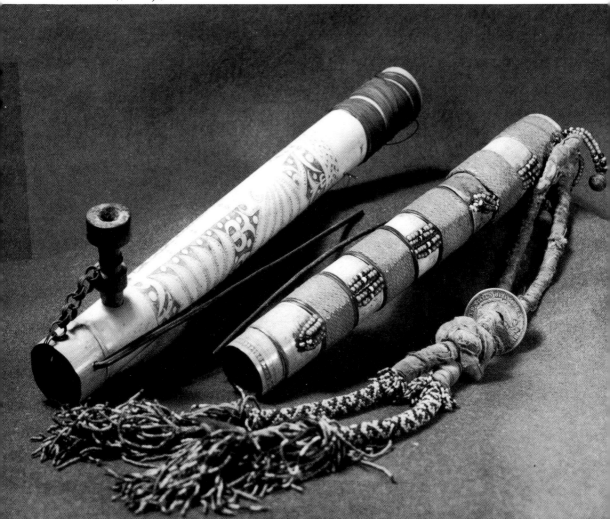

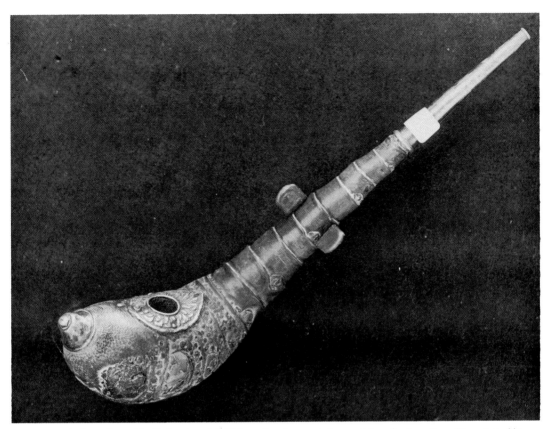

Originally from the Heide Collection, this Chinese bamboo root dry smoking pipe has silver fitments and is 8" long. Purchased in 1978 for $40. (Martin Friedman Collection) $100-200.

sometime in the Seventeenth Century, shortly after Japan and probably from the Japanese.[82] The dates are not as significant as the implements which were created to "drink" the smoke but it is worthy to note that Oriental smoking pipes thus date to the earliest clays of England.

There are three basic Oriental pipe styles: the dry tobacco, the water and the opium. The reader should understand, nevertheless, that although both the dry and the water were designed for tobacco, it was not uncommon to find that tobacco in these two varieties was laced with opium. The dry tobacco pipe of China and its dwarfish Japanese and Korean counterparts are essentially alike—a long slender pipe consisting of a bamboo or hard wood tube joined by a metal bowl on one end, a mouthpiece on the other. Although length is a relative term, I feel safe in saying that the Chinese dry pipe, when compared with the Japanese, but not necessarily with the Korean, is generally longer, has a larger bowl (perhaps twice the tobacco capacity) and its mouthpiece is rarely made out of metal. To establish specific and fixed lengths for each, as I had done with porcelain pipes, would be exceedingly foolhardy and dangerous—the wily and cunning Oriental is not bound by such conventions. More importantly, length was an important factor in the Japanese dry pipe since one made for a lady was almost always longer than one for a man.

The Japanese dry pipe, the kiseru, has a bowl shaped like an acorn, containing only a pinch or two of tobacco but its small size was overcompensated by frequent filling. "It is the custom of each smoker to roll the tobacco between his fingers into a ball of the exact size required to fit the bowl of the pipe, so that when turning

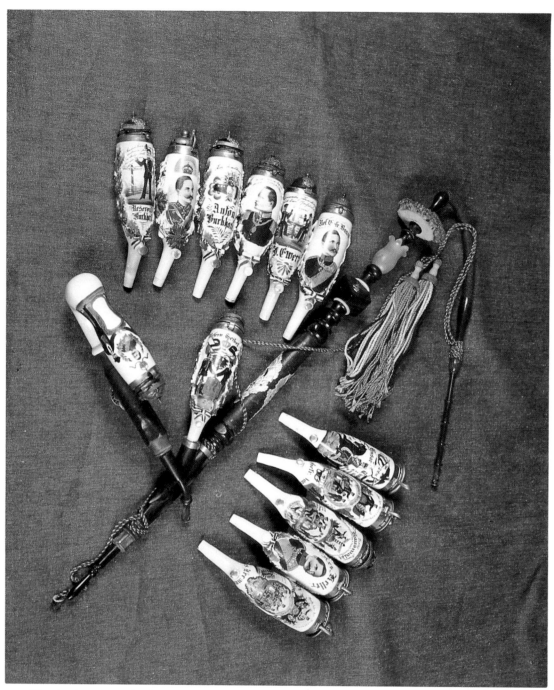

The classic regimentals of Imperial Germany are sought after not only by pipe collectors but also militaria buffs. The eleven bowls and two pipes displayed more than adequately indicate the colorful, varied and interesting history they retell. (Author's collection) $150-350.

Two bold relief wood pipes from Germany. The one on the left appeared on the cover of *Hobbies Magazine*, October, 1946 to accompany a feature story on the Heide Collection. It is carved pear wood with cherry stem sectioned by horn and horn mouthpiece. The theme on the bowl is Cupid inscribing the letter "L". A castle, in the background, is in low relief. The "L" is for Ludwig. The attribution, from Heide, is as follows: Ludwig I, King of Bavaria from 1825-1848, received this hunting pipe from his father, King Maximilian, when Ludwig was a student. He heired it to his son, Ludwig II, known later as the Mad King. Ludwig II distributed his father's effects, among them, this pipe which he gave to the head gardener. The gardener fled to America and Chicago. The pipe was used as rent money and eventually was purchased by Heide. The pipe was purchased by the author in 1978 for $100. Left: $200-600, right: $200-600.

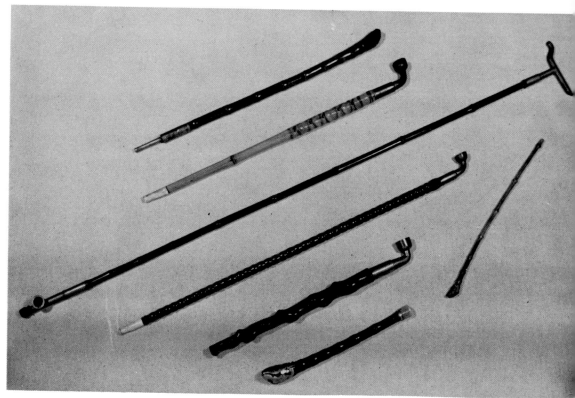

Assorted long dry smoking pipes from China and Formosa of various materials. Note that the bowl size is relatively bigger than the Japanese dry smoking pipes. (Dave Terry Collection) $75-250.

A melange of Chinese and Manchurian dry smoking pipes. From left to right: Silk handkerchief, tobacco pouch and pipe; a pipe set consisting of champlévé enameled pipe, pipe bag and tinder lighter pouch; ladies pipe set with pouch; embroidered pipe carrying case and pipe of ivory with paktong bowl and jadeite mouthpiece; in the lower right corner, a Northern Chinese incised pipe of brass with knocking keel; Chinese telescoping pipe in paktong. (Dave Terry Collection) $200-800.

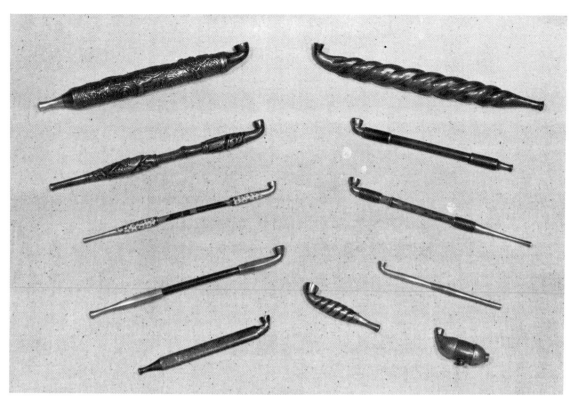

Japanese dry smoking pipes of brass, silver, paktong, cloisonné, iron and steel. The two on top are Sumo wrestler pipes of iron. The one in the lower right-hand corner is a miniature. (Dave Terry Collection) $100-800.

Japanese pipe sets. Upper left: bone case with leather pouch and inro; upper right: Daruma wood tobacco container, pipe in cloisonné and silver; lower left and lower right, pipes, leather carriers and pouches. (Dave Terry Collection) $300-1200.

Pipe from the Tinder Box International Pipe Collection, before and after cleaning. See Chapter 11. (Courtesy Doug Murphy) $150-350.

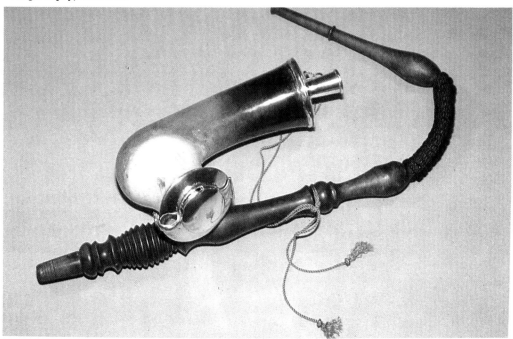

Opposite page:
Assorted wood pipes from the Jean-Charles Rhein Collection
1. Eighteenth century German boxwood pipe with inscription: *Amor and Psyche.*
2. Ulmer with horn stem from Budapest, around 1860.
3. Exotic Nineteenth century French wood and horn.
4. German, around 1800.
5. German, Eighteenth century.
6. German fruitwood, Eighteenth century.
(Courtesy Tabac Rhein) $250-900.

the pipe sideways to light it at the live charcoal it should not fall out; after every two or three whiffs a fresh ball is introduced."[83] According to Baron de Watteville then, the long, slender pipe and a small bowl would classify the Oriental as both lazy and economical! This long pipe was used by the gentry while a crude one-piece pipe of bamboo or root wood with a hollowed out bowl was used by the farmer, the peasant, the worker.

The material for the metal bowl was generally white copper, tootnague or paktong, a Chinese alloy of 50 parts copper, 31 parts zinc, and 19 parts nickel; other metals have included brass and black iron.[84] Bowls, shanks and metal mouthpieces have been ornamented with silver and gold inlay, surmounted with effigy and zodiac symbols, enamelled in champlévé and cloisonné. Tubes came in a variety of materials: mottled bamboo, ebony, root or lacquered hardwood. The Chinese, more than the Japanese, employed stone, jade, ivory and milk-white glass mouthpieces; the glass is, more precisely, translucent jadeite, a semi-precious stone of great beauty prized by the Chinese and considered sacred and magical. Jadeite mouthpieces have been found in a variety of colors: grass green, dark green, red, brown and cloudy. The traditional Japanese mouthpiece in a dry smoking pipe is of metal. The Korean version is almost the same, but Koreans were prone to make bowls of clay and wood. Solid ivory and jade pipes are often encountered as are solid metal pipes which evolved for the sumo wrestler.

The Oriental water pipe is more unique, compact and portable than the Near East version, is of metal and principally a Chinese expression. There are two general classifications: the Eighteenth Century variety was of bronze or brass, with long gooseneck stems emanating from trumpet-like figurals of the crane and elephant; other versions were plain, chased and bamboo finished obtuse cones.[85] The Nineteenth Century innovation was a box configuration from which flowed a metal stem. There is a receptacle for storing tobacco and a complete one would include a scraper with brush and pincers or tweezers for picking tobacco. "Water-pipes are also made of pure copper, and there is a great variety of shapes and designs. Those of tootnague are frequently inlaid with ornaments of copper, brass, bone, horn, tortoise-shell, or enamel. It would be easy to collect several hundred varieties in different parts of the country."[86] From Shanghai evolved a water pipe in the shape of a boot and eventually other styles appeared, designed to collapse or disassemble for portability. As to smoking it: "a servant fills with a little ball of tobacco, that he rolls with his fingers, the tiny bowl, which ends in a long tube reaching to the reservoir of water. Then he lights it with a paper match, and in one puff, the whole result of his toil goes off in smoke. He then removes the bowl, blows out the ash, and recommences the process."[87]

Perhaps the most maligned and misunderstood pipe is the opium, invented by the Chinese on Formosa in the Eighteenth Century, the only Chinese pipe which may employ a clay or porcelain bowl. The opium is maligned because of the drug, misunderstood because so many Orientalia collectors have labeled dry and water pipes as opiums. "The Opium-pipe consists of a stem from 40 to 50 centimetres long (about 22 inches), and as large as an ordinary flageolet, in bamboo or metal, according to the smoker's purse. The lower end of the stem has an opening in which the head of the pipe is fixed. This head is hollow, in shape round or cylindrical, generally in clay and sometimes in metal, and is fitted close to the top with

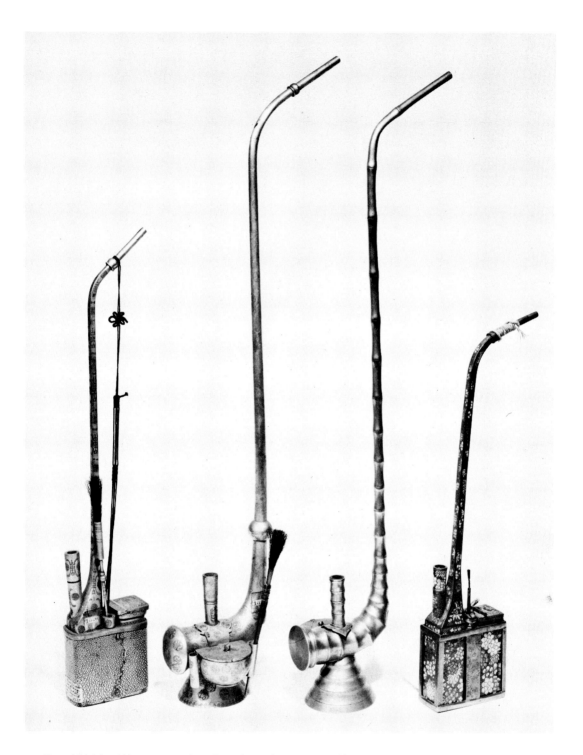

Lots 164-167, Chinese water pipes from the Parke-Bernet April 1978 auction. From left to right: Chased paktong, top of base and stem chased with stylized characters, base encased in green lizard, adjustable purple cord ($150); paktong with tobacco box side saddle, stem and body etched with ornamental medallions ($100); bamboo style paktong with tobacco box side saddle ($180) (Dave Terry Collection); cloisonné flowers and butterflies on ultramarine background ($300). (Copyright PB Eighty-Four, New York) $150-500.

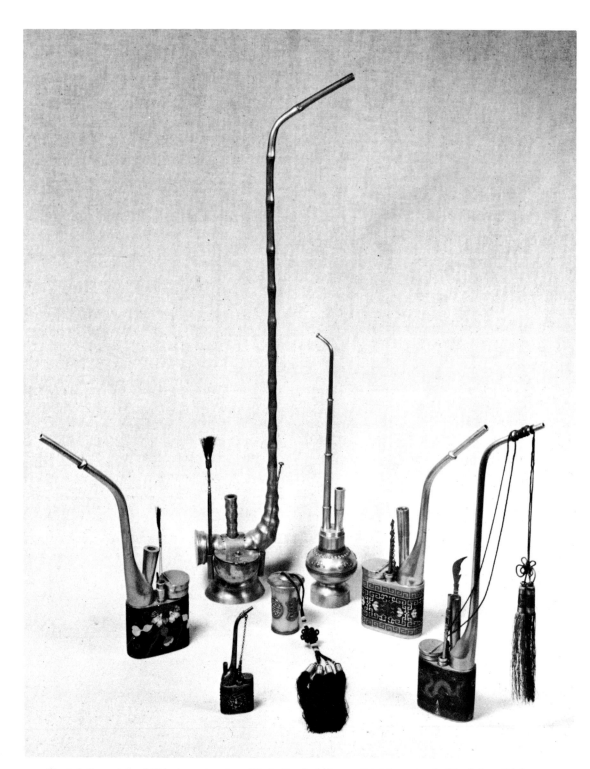

The infinite variety of Chinese water pipes. Illustration includes a miniature, several with cloisonné inlay and, at center front, a portable traveling companion. (Dave Terry Collection) $150-500.

This is a Heide item, an Indochinese Mandarin's water pipe, a Shibayama, of teakwood and inlaid with mother-of-pearl. It depicts a crane amidst rocks and plum blossoms, birds and Chinese characters. The ware tank and superstructure measure 8" high, 3½" in diameter. The mounts and bowl are of paktong and there is a silver chain and pipe cleaner. Purchased in 1978 for $475. (Charles P. Naumoff Collection) $300-800.

Above: a Chinese opium pipe in the form of a clenched left fist. A cylindrical bowl is inserted into the receptacle to make it complete. This choice collectable has an ivory hand and mouthpiece, mottled wood stem. And for comparison, below, a Chinese dry smoking pipe consisting of metal bowl, tortoise shell laminate stem and jade mouthpiece. (Courtesy le Musée du SEITA) Top: $300-500, bottom: $100-250.

a little shelf pierced with a small hole, on which the opium is placed and through which the smoke passes. For filling the pipe a long metal pin is used with which the smoker takes a very small quantity of the extract, which is a thick black paste like mastic varnish. Of this he makes a little ball as big as the head of a match which he holds near an alcohol lamp until it swells. As soon as this happens it is placed on the little shelf and lighted. The smoke is drawn slowly, and is swallowed for as long a time as possible. The pipe lasts for about a minute, and twenty to thirty aspirations are enough to empty it."[88] A shorter stemmed version of the Chinese opium is found in Borneo and New Guinea.

A doctor who witnessed opium smoking in Canton in October 1839 had this to say: "The apparatus, which was contained in a leather bag, consisted of an opium box, a pipe of peculiar construction, a lamp, & a steel bodkin about 6 inches in length. The opium, contained in a wooden box little larger than a lady's thimble, was a clear, dark, semi-fluid substance, resembling tar, or treacle, tho' of more consistence, & prepared, as I understood from the crude drug by boiling, straining & evaporating. The pipe, was of ebony, about 18 inches in length, & ¾ in diameter, had a brass bowl near the lower end, which was closed. The bowl resembled a pear, having its upper surface smooth & flattened, & a small aperture in its centre. Drawing a table, to the side of a bamboo couch upon which he seated himself crosslegged, our hero began by lighting the lamp, over which he placed a glass shade to render the flame strong and steady, and prevent it smoking. He then took a quantity of the drug the size of a pea on the bodkin's point, & held it a few seconds in the flame of the lamp, it swelled and took fire, emitting smoke of a strong aromatic & not unpleasant odour. Instantly blowing it out, he rolled it on the bowl of the pipe (by swiftly twirling the bodkin between his forefinger & thumb) then applied it to the flame & repeated the process 2 or 3 times. When sufficiently burned the bodkin was introduced through the aperture of the bowl, turning it round gently to detach the opium. Lastly, having made a deep expiration, he applied the pipe to his mouth, and its bowl to the flame, taking one long inspiration, by which the opium was almost entirely dissipated & converted into a dense smoke, this, was retained in the chest a short time, then emitted through his nostrils."[89] So fascinating and unusual was the rite of opium smoking and the prescribed ritual for the use of opium implements that the Stanford University Museum of Art has paid it special tribute in an exhibition entitled "Opium: Pipes, Prints and Paraphernalia," May 29–August 19, 1979.

The precious metals, the intricate design patterns of champlévé and cloisonné, the appliqué of tiny gems and stylized character chasing on all three classes of Oriental pipes have induced many a collector to specialize in one, two or all three of these Oriental luxuries which were prompted by a gift of nature—tobacco. Before 1766, the Tokugawa Shoguns of Japan who smoked owned long slender dry tobacco pipes with stems of silver and bowls of brass. In 1789, it was prohibited to use silver, gold or brass on these pipes, but the law must not have been enforced very strictly, for the legion of Oriental dry, water and opium pipes in the Parke-Bernet April 1978 auction were richly lavished with gold, silver, ivory, jade and cloisons.[90] Although the taste of Oriental metal pipes may be a bit unsatisfying, they have the unique artistic charm of the Orient, the articulation of the finest artisans and the value commensurate with any substantially carved meerschaum!

The implements of the opium smoker. From left to right: Chinese opium lamp with pewter base ($45); Chinese opium lamp in paktong from Peking ($35); Opium bowl stand with mother-of-pearl inlay and various opium bowls; three paktong opium containers and a diminutive opium lamp of brass, glass and horn. In the lower left hand corner are opium needles. Opium pipes from top to bottom: ovoid bamboo with stone bowl, paktong fitting, brass and copper inlay and jade ends; square lacquered wood with stone bowl; sectional ivory opium with the Eight Immortais in scrimshaw, pewter fittings and bowl in green nephrite; short opium pipe with clinched right fist and stone bowl. (Dave Terry Collection) $300-1000.

FOOTNOTES: CHAPTER 6

69. A Veteran of Smokedom, *op. cit.*, p. 69.
70. Fume, *op. cit.*, p. 131.
71. Forrester, Henry [Alfred Crowquill], A Few Words About Pipes, Smoking & Tobacco (Publication No 1, Arents Tobacco Collection, New York Public Library, New York, 1947) p. 74.
72. Cope's Tobacco Plant, Pipes and Meerschaum, Part the Second, the Pipes of Asia and Africa (Liverpool, 1893), p. 5.
73. Fairholt, *op. cit.*, p. 209.
74. Forrester, *op. cit.*, pp. 72-73.
75. An Old Smoker, *op. cit.*, p. 88.
76. Forrester, *op. cit.*, pp. 65-66.
77. Penn, *op. cit.*, p. 158.
78. *Idem.*
79. Fume, *op. cit.*, pp. 127-128.
80. Cope's, *op. cit.*, p. 23.
81. J. I. D. Hinds, The Use of Tobacco (Cumberland Presbyterian Publishing House, Nashville, 1882) p. 73.
82. Berthold Laufer, Tobacco and Its Use in Asia (Anthropology Leaflet 18, Field Museum of Natural History, Chicago, 1924) pp. 1-4; Shū Aoe, Satsu-Gu Tabako Roku (Zenschichi Maruya, Tokyo, 1881) p. 8.
83. Edward Vincent Heward, St. Nicotine of the Peace Pipe (George Routledge & Sons, Ltd., London, 1909) p. 129.
84. Bradford L. Rauschenberg, *Journal of Early Southern Decorative Arts* (May 1979), p. 70.
85. Laufer, *op. cit.*, p. 29.
86. Ibid., p. 30.
87. Cope's, *op. cit.*, p. 36.
88. Ibid., p. 37.
89. Forrester, *op. cit.*, pp. 6-8.
90. Aoe, *op. cit.*, pp. 9-10.

Chapter 7

ETHNOGRAPHICA II: AMERICA AND AFRICA

One Interpretation of a Calumet

"It is nothing else but a large Tobacco-Pipe made of Red, Black or White Marble: the Head is finely polished, and the Quill, which is commonly two feet and a half long, is made of a pretty strong Reed or Cane, adorn'd with Feathers of all Colours, interlac'd with Locks of Women's Hair. They tie to it Wings of the most curious Birds they find, which makes their Calumet not much unlike Mercury's Wand, or that Staff Ambassadors did formerly carry when they went to treat of Peace. . . However every Nation adorns the Calumet as they think according to their own Genius and the Birds they have in their Country."

Father Louis Hennepin, 1679

It would be very pretentious for me to believe I could impart in one chapter, however long, an adequate commentary on the myriad pipes of two diffuse continents, America and Africa. In these two land masses, the pipe has been native for hundreds of years. Unfortunately, American Indian and African pipes do not lend themselves to simple, stylistic definition and classification. These indigenous pipes evolved in a diffused and nonpatterned manner, as distinct expression and character of individual tribes and sects, not nation-states. As a result, there is no North or South American pipe, no African pipe, per se! Grouped together by common geography, the display would simply be a pipe melange of many peoples from that land mass—without visible traces or evidence of commonality. Apart, each pipe is an orphan requiring extensive investigation. . .and even unscientific conjecture. These two species defy the traditional and evolutionary developments cited in previous chapters.

A study of collectable American or African aboriginal and ethnological specimen pipes is, indeed, a major anthropological and morphological undertaking. My intention, therefore, is to provide an overview, set the stage and stimulate the necessary intellectual curiosity so that the reader will continue with his own independent research in this very fertile field of archaeological and ethnological pipe forms. Chapter 12 contributes to that end!

An introductory distinction must be made between archaeological and ethnological pipes. The former are aboriginal and classed as prehistoric; that is, pipes which belong to tribes or, better said, cultures prior to the presence of the White Man. Ethnological pipes fit into a class which evoke more modern influences.

North America was exclusively a land area of pipe smokers except for the Pueblos of the Southwest. Those native to northern and central South America and the West Indies were cigar smokers. The inhabitants of Central America and

Mexico were predominantly cigarette smokers, but extensive research of the Mayans has, of recent times, managed to prove otherwise. Indian pipes were to two general types—the straight or tube pipes in which the bowl and the stem were on the same plane or level and known throughout North America except for a few Eastern pockets. The right angle or elbow pipe has been found principally in the East, the Great Plains and in South America.

The Chippewas established their own peculiar classification:

    1. Women's pipes which were small, short stemmed and of minimal decoration.

    2. Men's pipes which were larger and better made than women's.

    3. Personal pipes of warriors with embellished stems in lengths up to five feet.

    4. Chiefs' and ceremonial pipes of more majestic quality than personal pipes.[91]

The pipes of North America were as varied as the materials from which they were made. The materials included both suitable and unsuitable substances: wood, bone, stone, antler, metal, chlorite, or steatite, stone coal, catlinite, limestone, trap-rock, gypsum, agate, hematite and pottery.[92]

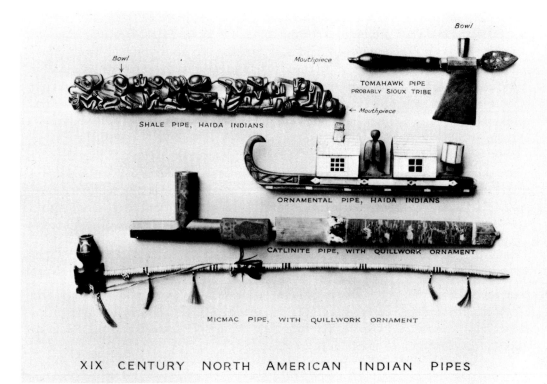

Courtesy British Museum

I defer to West's classification and cite, in capsule form, the most classic aboriginal pipe forms:

1. The tube is one of the oldest and most primitive, found practically in every aboriginal area of North America. Plain and ornamented, they took the shape of a today's straight cigar or cigarette holder. Some were fitted with mouthpieces of wood or bone.

2. The monitor is a flat or curved base upon which is mounted the bowl. Its form was likened to the steam-propelled armored warships once used for coastal defense in the United States and thus the name. As a general rule, the flat-based monitor is older and less adorned; the curved-based monitor or mound pipe is ornate and frequently surmounted with effigies and stylized characters. Monitors stemmed from the region of the Mississippi and Ohio Valleys and the Southeast.

3. The effigy is a genre which expressed an exceptionally artistic talent and includes the many stylized animals and birds into which the bowl was carved or drilled. Some exceedingly weighty effigy pipes encountered indicate that they were not intended for the use of one individual.

4. The disk is as the name implies. It is a pipe with a circular projection or disk on the shank which is perforated. Those found with rather long shanks bear the title handle disks. This pipe is rare and associated with the Iowa, Oto, Osage and Oneota cultures.

5. The handle is one with the bowl at right angles to the stem and a downnward extension in the vicinity of the bowl which acts as a grip or handle. The handle portion has been found in varying lengths and shapes.

6. The ovoid is simply a crude egg shaped devise, bored with two holes, one for the tobacco and one into which is inserted a stem or tube.

7. The vase is similar to the ovoid in principle but takes its name from the grooves and turns expressed on the exterior. It too requires a stem or tube.

8. The lens can be best explained as a flattened ovoid. It is a narrow circular bowl with bevelled edges. Again, this form dictates the use of a stem.

9. The micmac is an inverted acorn or vase-shaped bowl with a narrow neck and a keel or bar base. The bowl is decorated and often separated from the base by grooved lines or deep impressions. This pipe evolved among the Plains Ojibwa, Plains Cree, Blackfoot and the Micmac.

10. The bridegroom type can take any form or style. Its distinguishing feature is the presence of two stem holes!

11. The calumet, the peace pipe, the most revered possession in any Indian tribe, was the ceremonial pipe of catlinite, a red clay named after George Catlin. The Catlin letters described so vividly how the Red Man pledged his friends through its stem and bowl and that above all else, this pipe was to be placed in the grave alongside the tomahawk and war club. The calumet is principally from the Plains region.

12. The elbow or rectangular is often mistaken for the calumet, for it too is a right angle pipe but made principally of stone or pottery. The "T" shape, an elbow pipe with a projection forward of the bowl, is of this class.

Pipe bowls from the Great Lakes. Upper left: Wyandot black stone, recumbent holding barrel and glass, 19th century; upper right, Ojibwa figural head, c. 1860; next under, Eastern Woodlands "T" shape elbow in brown stone, bowl carved as a human head, c. 1850; center left, Eastern Woodlands green stone of human wearing a European style hat, c. 1840; lower left, Ojibwa black stone bowl, c. 1850; lower right, Ojibwa wood bowl, inlaid with lead and varnished, c. 1860. (Reproduced by Courtesy of the Trustees of the British Museum) $200-1200.

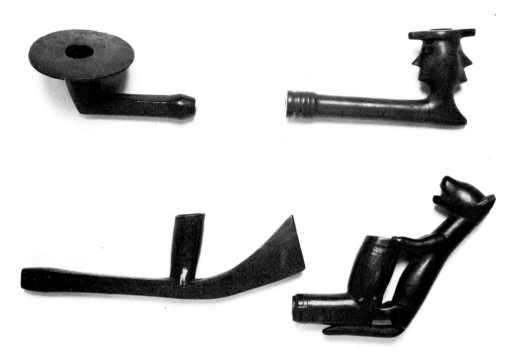

Eastern North American Indian pipe bowls. Upper left: black stone disc, Eastern Woodlands, 18th century; upper right, green stone double-headed figural, Eastern Woodlands, 17th century; lower left, Cherokee, black stone "T" shape elbow, 18th century; lower right, black stone effigy elbow, Eastern Woodlands, 18th century. (Reproduced by Courtesy of the Trustees of the British Museum) $200-1000.

Pipes of the Plains. Calumets and elbow pipes in stone and iron. (Courtesy US Tobacco Company Museum) $400-1500.

13. Iroquoian pottery stone is not easily confused with any other aboriginal shape. It is a pronounced trumpet, with and without decoration. Other pipes attributed to the Iroquois include right angle, stone ovoid, vase and lens.

14. The pottery can be anything from crude to embellished, exclusive of the Northern Iroquois trumpet-like pottery.

15. The pebble is an improvisation of a concretion or geode with a convenient cavity. All that is required is the boring of a stem hole.

16. The Northwest Coast totemic is a highly decorative effigy pipe, full of action, activity and anecdote. Made of argillite, slate or stone, it was elaborate, aesthetic and inventive.

17. The Eskimo with its Asiatic influence made of ivory, stone or metal.

18. The tomahawk made first by the White Man.

Pipe stems are more difficult to classify. Generally, they were straight, of ash, sumac and other soft woods, detachable from the bowl and of varying lengths. Another dimension is shape and both flat and round stems were popular. Decoration included porcupine quills, fur, feathers, beadwork, horse and human hair, fringed trade cloth, skin and an endless array of incised and burnt carvings, geometric designs and stylized characters.

Wooden pipe bowl from the Northwest Coast of America. Wolf inlaid with Haliotis shell, 19th century. (The Oldman Collection, Courtesy British Museum) $300-1500.

156

Northwest Coast Haida argillite pipes, 1840-1860, carved with traditional subjects. Above: European-Americans, house and dog, inlaid with bone and ivory; center: European-Americans, dog and house; bottom: European floral motifs, thistles and oak leaves, people and a sea monster (?). (Reproduced by Courtesy of Trustees of the British Museum) $1000-3000.

Three Haida argillite pipes portraying mythological subjects. Each measures between 15-20 inches in length, c. 1850. (Reproduced by Courtesy of the Trustees of the British Museum) $1500-4000.

Wooden pipe with mythological figures. This North American pipe has a copper tobacco cavity and Haliotis shell and ivory inlay, circa 1830-1850. (The Oldman Collection, Courtesy British Museum) $500-2000.

An Eskimo pipe of walrus tusk. (Courtesy Wills Collection of Tobacco Antiquities, Bristol, England) $1000-3000.

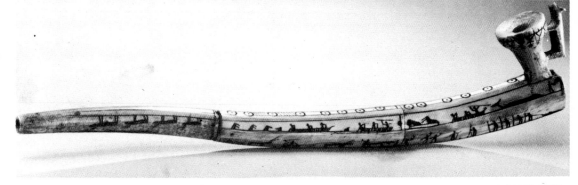

Eskimo pipe, detachable bowl with effigy dog surmounted. (Courtesy US Tobacco Company Museum) $1000-3000.

A word or two about the much prized calumet. Fundamental to its understanding is that the word comes from the Latin calamus meaning reed, stem or shaft. The calumet, the stem, was ceremoniously symbolic independent of the pipe bowl of catlinite! Thus, "...when the pipe became an alter, by its employment for burning sacrificial tobacco to the gods, convenience and convention united the already highly symbolic calumet shafts and the sacrificial tobacco altar, the pipe bowl; hence it became one of the most profoundly sacred objects known to the Indians of northern America."[93] There were calumets for every kind of public event, engagement, occasion and commerce between Red Men and between Red Man and White Man. It was in universal use and many different tribes visited the quarries of Southeastern Minnesota to extract this fine-grained claystone with the peculiar qualities of softness and firmness. The evolution of this pipe is fascinating. Older specimens like the smooth and unornamented calumet presented to the Jesuit Father Marquette by the Sioux were supplanted by the more decorative ones used in later English encounters with Indians. For the want of a common language, wampum and pipe were symbols of treaty and contract; eventually, the decorative wampum belt influenced the decoration of peace pipe stems.[94] With the passage of time, the calumet has come to signify the entire pipe.

At the other end of the socio-cultural spectrum was the tomahawk pipe or pipe tomahawk. Since the hatchet and axe were early European and American articles of trade and since a pipe was used to celebrate pacts of peace, the tomahawk pipe was a natural mix by European traders of the peace pipe and "burying the hatchet." From about 1780–1900, tomahawk pipes were given in trade and as mementos until Indians themselves were equipped to make them during the reservation era. The earliest English examples were of steel or brass with a steel edge. As its popularity waxed, the Nineteenth Century saw the introduction of elaborate presentation tomahawks with gold and silver inlay, less a cutting instrument, more a symbol of authority and power. Blade and bowl styles varied with the nation. In the English version, the smoking bowl was round, the blade long and narrow; a further refinement was the half-axe where the forward edge is perpendicular to the handle. The French or spontoon style takes the form of a kite, often with and sometimes without curled flanges on each side. The Spanish style was always the most ornate; it had a more tubular bowl and a blade which was narrow at the shaft to very wide at the cutting edge. The tomahawk pipe eventually transitioned into a souvenir...in both steel and stone with embellished stems—an overarticulation of what once began as a symbol of peace and trade.

The English, French and Dutch clay trade pipe is a category which, for the collector, can be classified as either North American ethnographica or would fit comfortably in a clay collection since the clays were originally specimens of continental manufacture. When the colonists began to produce tobacco in the Seventeenth Century, they made pipes concurrently for Indian trade. On September 10, 1677, 120 trade pipes were part of the exchange for the land purchase between Rankokas Creek and Timber Creek, New Jersey; in 1682, William Penn's land purchase included 300 tobacco pipes in the exchange.[95] By 1692, the Indians were introduced to various other trade pipes: wood, tin and wampum.

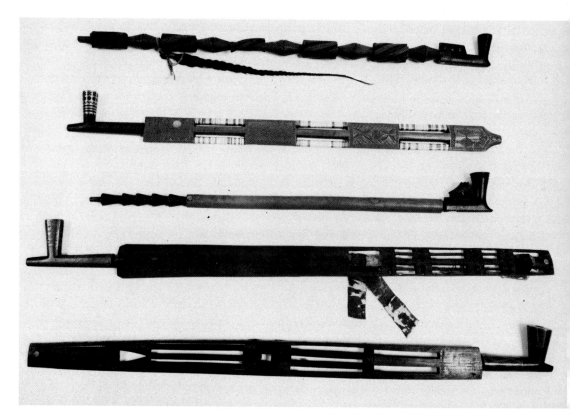

Calumets from the Great Lakes and Eastern Plains. From the top: black stone bowl, wood stem, decorated with scalp lock and silk ribbon, Ojibwa, c. 1840; catlinite bowl inlaid with lead, wood puzzle stem, painted red and green, wrapped with blue, orange and white quillwork, Eastern Sioux, c. 1850; black stone bowl, wood stem finished in cone work, Ojibwa, c. 1850; catlinite bowl inlaid with lead, wood puzzle stem, painted green, green and blue silk ribbons, Ojibwa, c. 1840; catlinite bowl inlaid with lead, wood puzzle stem, Ojibwa c. 1840. (Reproduced by Courtesy of the Trustees of the British Museum) $500-2000.

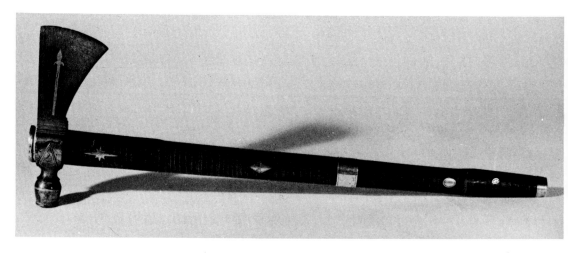

English half-axe, late 19th century. Ornamentation is refined and evinces machining. (Courtesy Wills Collection of Tobacco Antiquities, Bristol, England) $500-2500.

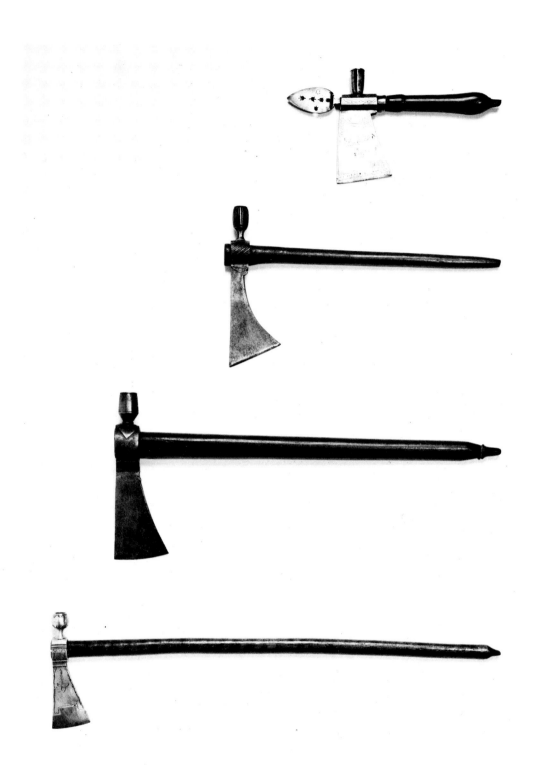

Eastern Woodlands pipe tomahawks from the period 1780-1830. Top to bottom: screw fitted bowl and spear point, blade engraved with Indian intimidating a white man and a coat of arms on obverse; a plain half-axe; a half-axe of British manufacture, attributed to the Hudson's Bay Company; engraved brass blade with steel edge, English manufacture of a type used in the Revolutionary War. (Reproduced by Courtesy of the Trustees of the British Museum) $500-2500.

A variety of tomahawks. From top to bottom: Spanish, steel and wood, first half of the 19th century; French spontoon in steel and wood with tear drop cut out and flanges, 1860-1880; cast iron, possibly of a home industry, Nebraska, after 1900; speckled catlinite, English half-axe style; spontoon pipe bowl, in catlinite, of Sioux manufacture for trade and tourism. (Courtesy US Tobacco Company Museum) $500-2500.

In South America far less is known, but primitive smoking implements of stone, cane and pottery have revealed the predominant forms of the tubular pipe and widely diffused elbow prototypes in Chile, Venezuela and Argentina. Whether for the pleasure of smoking or for the cure of pain, there is ample and unquestionable evidence of tobacco use from at least the Sixteenth Century. As snuff, as incense, as medicine, as a crude cigar and in the pipe, tobacco played an important role in South America with a concomitant development of diffused pipe styles in various mediums.

The Continent of Africa is considered a region where rather than as a pleasurable experience or amusement, the aims of smoking have been one of intoxication and another of a rite connected with public observances, rank consciousness and social etiquette.

The endless aboriginal and picturesque pipes of uninhibited Dark Africa again represent a panoply of makeshift, crude, rustic and even elaborate styles in wood, bone, gourd, coconut, horn, ivory, clay, bronze, iron, copper, brass and base metals. And in size, anything from a bowl which could hold several handfuls of tobacco to dainty petite bowls has been discovered. But the diffusion has made generalizations about African smoking pipes difficult; originality, in pipes, is the essence of Africa!

South American pottery. (Courtesy US Tobacco Company Museum) $300-700.

A 20th century black pottery pipe from Uganda. (Courtesy Alfred Dunhill Ltd) $100-250.

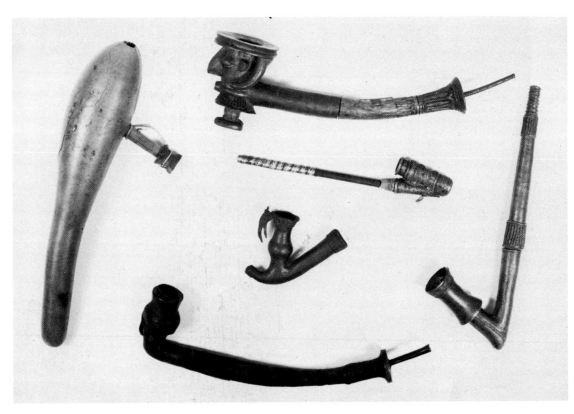

Assorted Africana of brass, wood, gourd and iron from various collectors. $100-500.

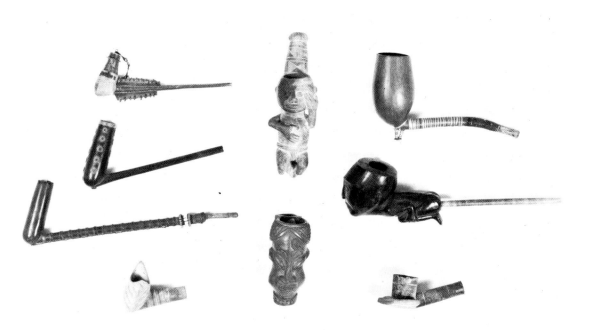

Mixed media of Africa, once more. Wood, pottery and metal. $100-500.

In Northern Africa, the proximity to other Mediterranean influences has kept the hubble-bubble alive! In the Nineteenth Century, Dahomey pipes were very expressive—reddish yellow clay effigy bowls of birds, fish, canoes and humans! In the Congo, there are traditional European forms, gourd water-pipes and bamboo pipes; many are fringed, as the calumet, with hair and beads. The kaffir pipes of South Africa have been likened to Dutch trade clays which the natives had seen during the days of Dutch exploration. Only the form of the kaffir can be traced back to Europe for the African expression has been found in stone, wood, metal and pottery. The Moorish women who lived in the Sahara smoked "gold and silver pipes, whose short stems were adorned with coral and amber."[96] The natives of Guinea had long pipes up to six feet in length. Among the Ashantees, men carried gold pipes, while their women carried clays. Here also, "the pipe-bowl represents some queer animal, human or bestial, and the long flexible reed tube, conducing to cool and clean smoking, is tastefully adorned with silver wire."[97] In Lagos, on Africa's West Coast, an original but certainly awkwardly attractive pipe is found—dark clay twin and triple bowled pipes.

According to Paul Gebauer, once a missionary in Cameroon for thirty years and now a special lecturer in anthropology at Linfield College, McMinnville, Oregon, Cameroon is the ideal collector world where clay, terracotta, wood, ivory and metal, in the hands of inspired craftsmen, have become stylized pipe bowls portraying a panoply of zoomorphic fantasy animals, miniature masks and head-gear while pipe stems are carved like house posts! Pipes from this small country, claims Gebauer, are themselves a study, the incorporation of art, folklore, mythology, history and external influences into expressions of a grassland culture mingled with the splendor of early feudal rule.[98] Traditionally, wood stems are

Brass studded gourd pipe from Angola. (Courtesy US Tobacco Company Museum) $300-800.

Wooden effigy pipe, Basuto tribe, Africa. (Courtesy Wills Collection of Tobacco Antiquities, Bristol, England) $100-300.

Effigy pipe of wood and metal from Angola. (Courtesy Wills Collection of Tobacco Antiquities, Bristol, England) $300-800.

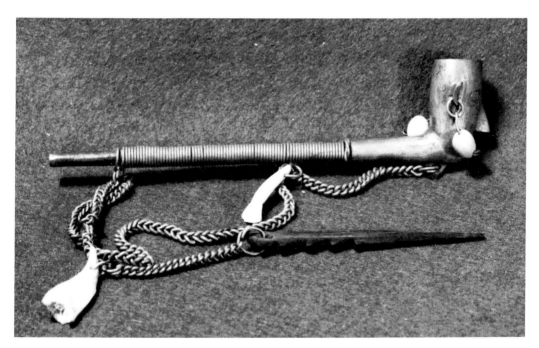

A 5½" African figural in brass. Ornamentation in copper wire and animal teeth. Note sophisticated retaining chain and probe. (Ken Erickson Collection) $100-200.

African pottery bowls from Cameroon and Bali. Second from the left and second from the right are from the Tinder Box International Pipe Collection. Santa Monica, CA. Others from the Pozito Collection, Madrid. $300-1000.

inlaid with copper wire, plated in brass and tin, studded with brass tacks while carvings have included the python, spider, leopard, bat, lizard, crocodile, frog and cowrie shell as well as fertility and long life symbols, geometric designs and bead work. "Fashion and historic events had an impact on pipe art. Pipes are records of tastes and change. The ceremonial pipe of one chieftain, for instance, depicts a soldier of the German Colonial Army, rifle and all. This style was popular after the military conquest of Morgan, Stetten and Dominick in the 1890's. With the coming of British rule and law, the German soldier was replaced by the figure of a judge sitting in session, stroking his beard or smoking his own pipe."[99] As elsewhere in Africa, Cameroon pipes are symbols of rank and authority with pipe size and decoration commensurate with a person's social stature and wealth. Women, the heaviest smokers, are accustomed to small pipes but those of royalty and rank do defy this tradition by touting highly ornamented and man-sized pipes!

While clay and wood pipes have become rare Cameroon manufactures, pipes of brass and bronze are still being made today at two Bamum casting centers, Fumban in the East and Nkwen in the West.[100] In clay, few craftsmen remain but in at least one Cameroon village, Bamessing (now Nsei), noted for its fine quality clay pipes since before 1880, the methods of manufacture remain unchanged today—iron and raffia tools for sculpting and the processes of sun-baking, grass-fire hardening and liquid sprinkling for color variations.[101] In the last decade, a concerted effort by the Peace Corps and the Cameroon Government has resulted in a rejuvenation of pipe crafting by traditional ways, making it difficult to discern between the true artifact and a national enterprise which has been rekindled.

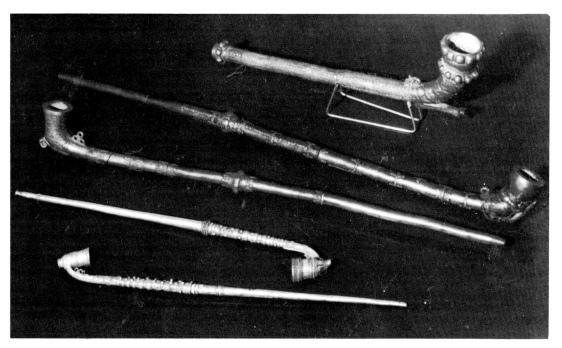

Cameroon brass pipes. (Courtesy Austria Tabakwerke, Vienna) $200-800.

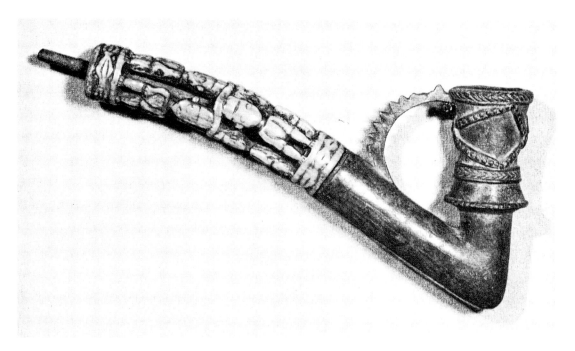

A bronze and ivory pipe of the Bamun, Cameroon. (Courtesy Africa Museum, Meerwijk Berg en Dal) $300-1000.

Fetish water pipe from the Congo. $300-1000.

Chieftans' pipes. On the left, brass, decorated with waterbucks' heads in low relief, a human head and a tiger's head. On the stem are six moulded water buffalo heads. Overall length is 43½". Purchased in 1978 for $225 (Charles P. Naumoff Collection). On the right, another brass bowl, with three stylized bird heads. The cover is surmounted by a seated monkey; a coiled snake at the mouthpiece confronts the smoker. Purchased in 1978 for $225. (Martin Friedman Collection) (Copyright, PB Eighty-Four, New York) $300-1500.

I would hope that one day, the inquisitive student of African art and the avid collector of ethnographic pipes join forces to write the ethnohistory of Sub-Saharan pipes. To the best of my bibliographic knowledge, an extensive study is still wanting! There is ample documentation on African tribal masks, weapons and articles of fetishism, yet no authoritative source book for the treasure-trove of African relics related to the socio-cultural phenomena of tobacco and smoking. Without one, there will remain the ever-present difficulty of classification, attribution and evaluation of pipes from the genus Africa!

FOOTNOTES: CHAPTER 7

91. Ralph Linton, Use of Tobacco among North American Indians (Anthropology Leaflet 15, Field Museum of Natural History, Chicago, 1924) pp. 13-14.
92. George A. West, Tobacco, Pipes and Smoking Customs of the American Indians, Part I, Text (Greenwood Press, Westport, Connecticut, 1970) p. 382; Joseph D. McGuire, Pipes and Smoking Customs of the American Aborigines, Based on Material in the U.S. National Museum, Annual Report of the U.S. National Museum, 1896-97 (Government Printing Office, Washington, 1899) p. 626.
93. Linton, *op. cit.*, p. 24.
94. McGuire, *op. cit.*, pp. 631-632.
95. Ibid., p. 461.
96. Cope's, *op. cit.*, p. 53.
97. Ibid., p. 54.
98. Paul Gebauer, "Cameroon Tobacco Pipes" *African Arts,* Winter 1972, p. 31.
99. Ibid., p. 33.
100. *Idem.*
101. Ibid., pp. 34-35.

Chapter 8

METAL MISCELLANEA AND COGS

"Here's to the Hookah with snake of
   five feet,
Or the 'portable' fix'd to one's 'topper;'
Here's to the Meerschaum more naughty than
   neat,
And here's to ALL PIPES that are proper.
Fill them up tight
Give 'em a light;
I'll wager a smoke will set everything right."

>   From A Song, After Sheridan
>   February 1875

Of the many pipes described in this book, the one substance I have never understood, and more so after having smoked from it, is metal! Yet pipes of metal were fashionable in Europe, were predominant in the Orient and Africa and can be the most exquisite and best executed objet d'art in a collection.

The first written acknowledgement of occidental metal pipes in use was in 1595: ". . .I think we shall not need to thinke our earthen or siluer pipes more vnapt than those which the Indians made of Palme leaues and such like."[102] Metal pipes had the advantage of unbreakability when compared to the infant English clay pipe industry and were "designed for travelers and huntsmen, for whom the clay pipe was too fragile."[103] There are examples of sectional silver pipes from the mid-Seventeenth Century which either disassembled and fit into a carrying case or telescoped for easy portability. Churchwarden facsimiles were fashioned in sheet and cast iron in the Eighteenth Century. Henry René d'Allemagne's 1924 catalogue entitled DECORATIVE ANTIQUE IRONWORK, the ulimate pictorial treasury of ironware, illustrated European chased iron and steel pipes fabricated for use in Europe. There is the long-stemmed conical bowled iron pipe known as the traite or traffic pipe, named so as the popular smoking apparatus of the European slave merchants. Somewhere in America today, Miles Standish's little iron pipe, in the likeness of a small clay, is still being preserved. And, at one excavation at Jamestown, Virginia, the fragments of what appears to have been a pipe of pewter were unearthed. For a short time in Italy around 1900, wood-lined cast-copper pipe bowls of the Debrecen and Ragoczi genre were being manufactured for the smoking public and brass pipes are in the possession of the Belfast Museum.

As a general appraisal, nevertheless, metal pipes were not in wide use in the Western World, never achieved acceptability and were even considered dangerous weapons. "Metallic pipes, made of silver, brass, or iron, are now going out of fashion, but are of national use in Thibet. They are entirely impracticable, being easily

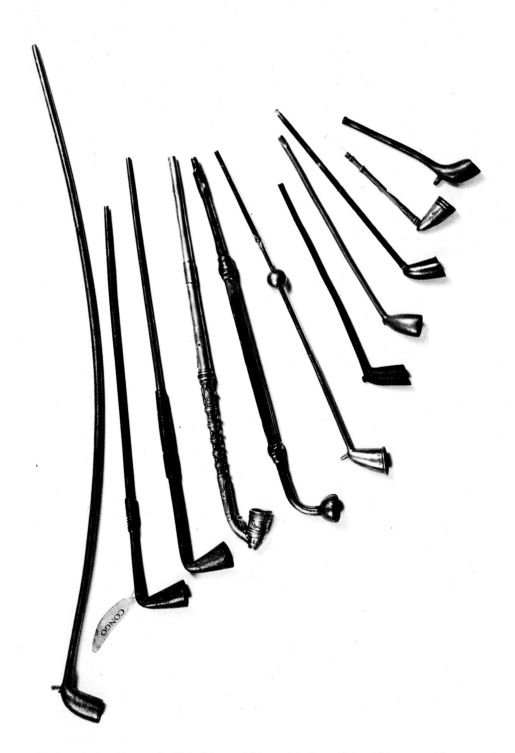

A mix of metals from Europe, the United States, Africa and the East, 16th to 19th century. The one on the bottom, an 18th century English iron churchwarden, belongs to the publisher, Peter B. Schiffer. All others are in the US Tobacco Company Museum. Note the two traite or slave merchant iron pipes above the English churchwarden. (Courtesy US Tobacco Museum) $100-700.

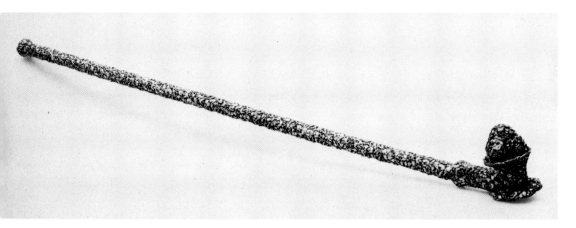

This is a premier chibouque! It is a 37" silver presentation pipe, embossed with grapevine design and accented by small pearls, rubies and turquoise. Purchased in 1973 for $220. (Charles P. Naumoff Collection) $500-1000.

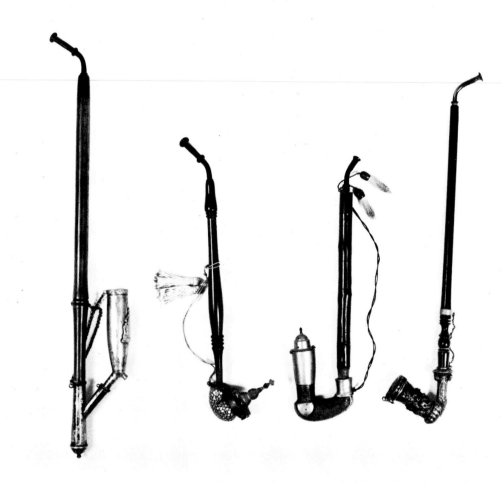

Nineteenth century metal pipes in a variety of styles. Left, metal gesteckpfeife in silver plate: mother-of-pearl on brass with obelisk dome, probably a Western European export; another for export with sheet metal mountings on wood bowl, inlaid with copper wire; bas-relief sculptured sterling silver bowl from Holland with shank surmounted high relief heads, ebony stem. $200-800.

Top: The legendary Greek warrior is depicted in this tasteful French 17th century silver pipette. A hand probe or pricker is attached for convenient cleaning. (Courtesy le Musée du SEITA) $200-500.

Bottom: An 18th century French silver pipe with a rather unique reservoir. This is the Roman goddess Minerva, the Greek goddess Athena. As either, she is wisdom and the Arts! (Courtesy le Musée du SEITA) $200-500.

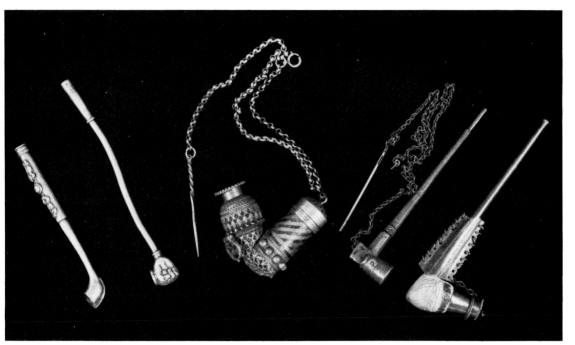

Metal-varia. Left to right: two miniature paktong opium pipes (bowls are missing): a brass turned and filigreed copoq, engraved **Constantinople, 1905** with pricker and retaining chain: figural brass Igorot pipe with pricker and retaining chain from Philippines: clay bowled Kaffir pipe with brass fittings and windcap. (Author's collection) $100-200.

Silver pipe of an eagle's talon holding a cup, of unknown provenance. The pipe measures 5¼" long and has briar inset. (Charles P. Naumoff Collection) $100-300.

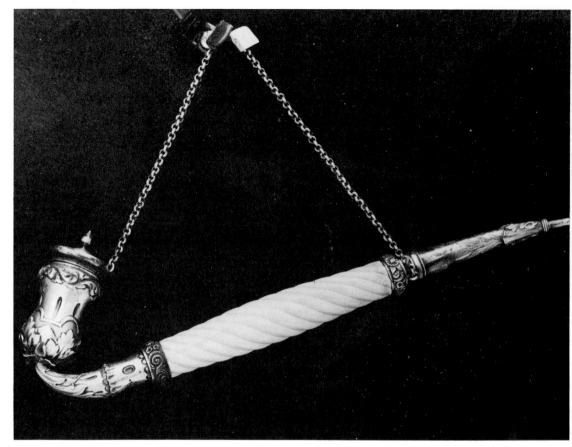

Heide purchased this silver and ivory gem in 1914 in Reading, Pennsylvania. Made by Tiffany of New York in 1890, it was a gift from the Lord Chief Justice of England, Richard Everard Webster, to Jacob M. Gordin, a Russian Jewish playwright living in America. This elegant 8" pipe was purchased for $60 in 1978. (Martin Friedman Collection) $1000-2500.

heated, non-absorbent, and conveying the smoke hot enough to burn the tongue."[104] Discounting the minimal appeal to a smoker, except in the Orient and Africa, metal pipes deserve some recognition for they play a significant role in the miscellany of a collection.

Now a brief encounter with the COGs, an acronym I coined as an umbrella word for the pipe-ana which include curiosities, oddities and gimmicks. There are a multitude of pipes befitting this designation but I have included only a sampling. If your collectable has not found a home yet, perhaps it belongs here!

Let me first set aside the mystery of glass. Glass pipes are rarities, but like Staffordshire, Whieldonware and Prattware, these too were impractical to smoke but quite decorative; they have adorned collections and have been displayed in tobacco shops. Most are mid-Nineteenth Century from either English glass bottle factories at Stourbridge, Bristol, Nailsea or from Venice. Sizes and colors abound, a tulip-shaped bowl most common to all. As to construction, they have been found as single knobbed pipes or sectional pipes interconnected by glass friction sockets. In any form, they, as puzzle pipes, are the proverbial conversation pieces.

Infrequently, a collector will occasion upon a solid amber pipe! This skeleton-in-hand measures 8½" long and 3½" high. The delicate cuffed female hand is finished with an 18K rococo scrolled band and amber stem. Purchased in 1976 for $275. (J.E. Opperman Collection) $750-2000.

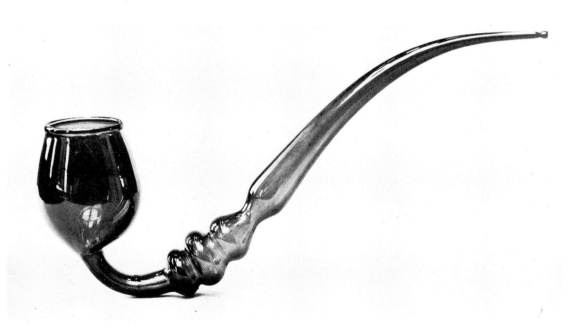

Bristol translucent cranberry color glass pipe. (Courtesy Wills Collection of Tobacco Antiquities, Bristol, England) $250-600.

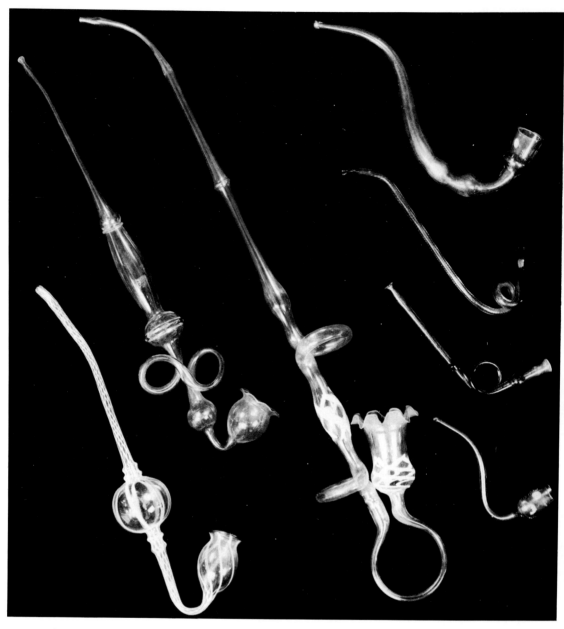

Venetian art glass pipes—fragile flaired decorative pipes. (Courtesy US Tobacco Company Museum) $100-200.

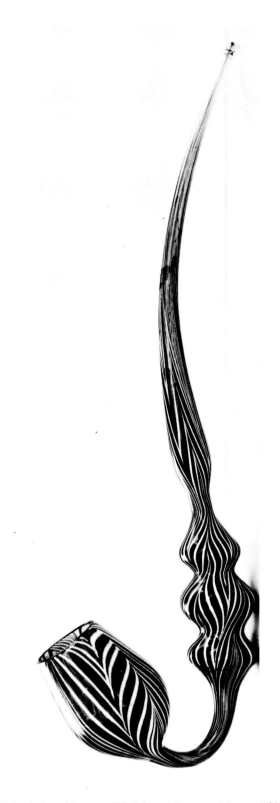

A veined example of Bristol glass. (Courtesy US Tobacco Company Museum) $250-600.

On the Walcheren Islands, Zeeland Province, lives a hearty breed of Dutch fisherman. As most Dutch, smoking is a national past-time even at sea. It is said that the inhabitants of that island developed the first nose warmer pipes, miniature delicacies crafted of lightweight metal with horn or hard rubber stems, sea-going pipes to brave the chill, early morning salt-sea air. Nose warmers were extremely short and when filled with tobacco, the smoke rose from these minis toward or into the nostrils. Inhaled, the smoke brought warmth for the ever-so-few pinches of tobacco this pipe held. I've not encountered many, but they are, indeed, smokable curiosities.

And since these nose warmers are of the mini class, a brief aside to mini pipes. As with any miniature collectable, it is difficult to separate between a salesman's sample, a toy or an actual pipe when under five inches in length. In any case, minis can be an exciting class of curiosa. A collector will often encounter many in this dimension: opera, between-the-acts, ladies' or after-dinner pipes; these are precise titles for such charming well-executed delights which hold perhaps one or two minutes of genteel puffing for a theatre intermission or to accompany an after-dinner cordial. They are legitimate pipes, daintier certainly than their standard size brethren, but nevertheless smoking utensils.

Another unconventional pipe mentioned by Aschenbrenner, Dunhill and Fairholt is the shell. In Normandy and Brittany, the channel provinces where Frenchmen even today associate themselves with these pre-1790 départements, the fishermen craft indigenous crustacean pipes. These are also native to Japan, Australia and the Philippines. The fitments are usually of metal and the cephalopoda, murex and naticidae are lined with meerschaum or contain a meerschaum bowl inset. A push stem is all that is required to get these sea urchins going! Across the Channel, natives of Cornwall have been smoking a cruder version—the unadorned and humble crab or lobster claw.

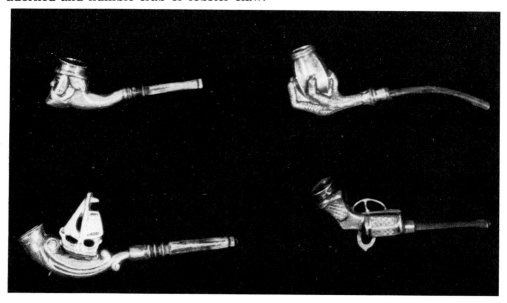

Nose warmers of the Walcheren Islands. The two on the left are in the Pozito Collection, Madrid; the two on the right are part of the Tinder Box International Pipe Collection, Santa Monica, CA. All measure less than 4½". $50-150.

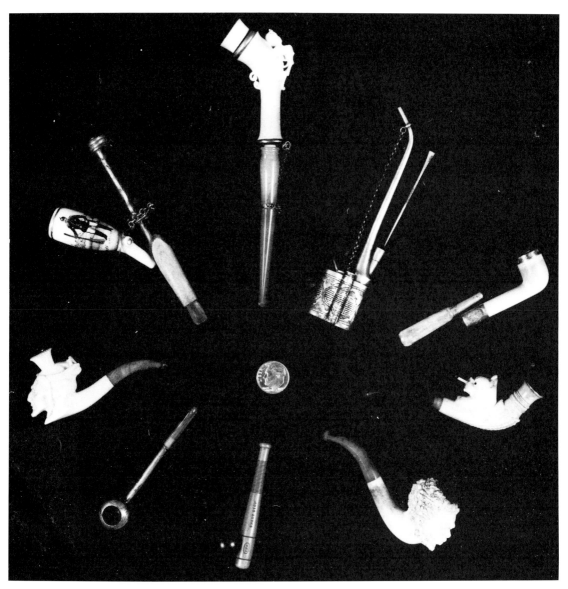

Miniatures from everywhere! Starting at 12 o'clock, a cigarillo holder in meerschaum and amber with full relief cherub; miniature Chinese water pipe in chased silver; between the acts pipette in meerschaum, amber and silver; cigarillo holder with surmounted dog smoking a cigarette; miniature nubian in meerschaum; GBD short stop; silver and amber opera pipe; miniature devil in meerschaum; miniature regimental with porcelain bowl and cherry stem. (Author's collection) $50-250.

Shell pipes from Brittany, Normandy and Cornwall. In the upper row are two from the Pozito Collection, Madrid; the two on the bottom, Tinder Box International Pipe Collection. $150-450.

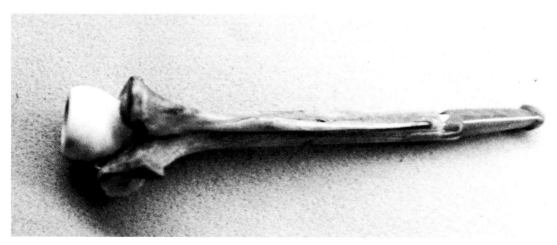

A pipe made of bird's leg with meerschaum inset. (Metromedia Collection, Duke University Art Museum, Durham, NC) $100-250.

An elk horn pipe from Northern Europe, 19th century, 10 inches in length, with low relief carving of lyre player. (Courtesy Duke University Art Museum, Durham, NC) $200-700.

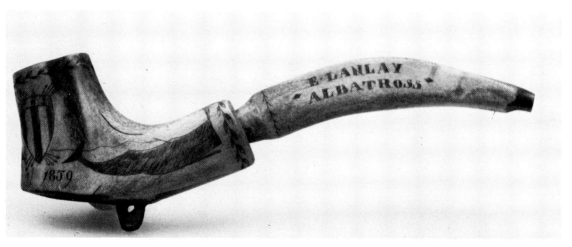

A whaler's scrimshaw bone pipe. One side features a harpooned whale spouting water, the date 1859, the name **E. LANLAY** and what most likely is the name of a ship, **Albatross.** On the other side are scrimshaw carvings of whales, hunters and birds. Purchased in 1974 for $350. (Charles P. Naumoff Collection) $250-600.

Pipes of ivory are not unknown. They originate in Alaska and Siberia from walrus or mammoth tusk. There is also the handmade scrimshaw variety carved by whalers in ivory and in bone. The most precision execution of ivory pipes was that of Dieppe, France. As early as 1612, the artisans of Dieppe had established a reputation for carving bibelots in ivory, horn and tortoise shell. In the Eighteenth Century, Dieppe sculptors already rivaled, in ivory, the minutest details of the best lace makers. One distinct pipe which evolved was a majestic pipe made entirely of ivory with intricate perforations, rococo and serpentine scrolls intertwined with grape and leaf foliates surrounding the bust of some mythological female in ornate headdress and lace collar; at the base, the head of a griffon or other such allegorical winged beast.

Antler and staghorn pipes are interestingly rustic. And the gourd pipe, the calabash of Mark Twain and Sherlock Holmes, can be a popular collectable. The Dutch observed Cape Town natives smoking hemp in gourd pipes as early as 1612. This graceful cucurbitaceous plant can be grown at will and whim since pegs or stakes implanted around the stem and body direct its growth. Turning the gourd frequently aids in ripening and even color, but no two ever develop alike. Gutting, seeding, polishing and drying have resulted in some beautiful African specimens in natural shapes or were crafted into the saxophone Sherlockian style, surmounted with meerschaum or clay insets, silver fitments and fine quality stems, as were manufactured early in this century.

A Dieppe ivory in all its regal majesty. (Courtesy H. F. and Ph. F. Reemtsma, Hamburg). A Dieppe replete with ornate beaded stem is in the Irving Landerman Collection. $500-3000.

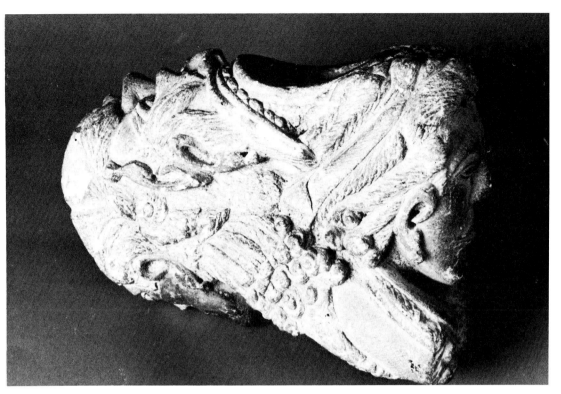

A pipe bowl of sculptured white stone, France, 19th century. In the round, facing right, the bust of a woman; underneath, the torso of another woman; above left, a mustached man. The bowl is rimmed by a mouth agape. (Courtesy le Musée du SEITA) $300-500.

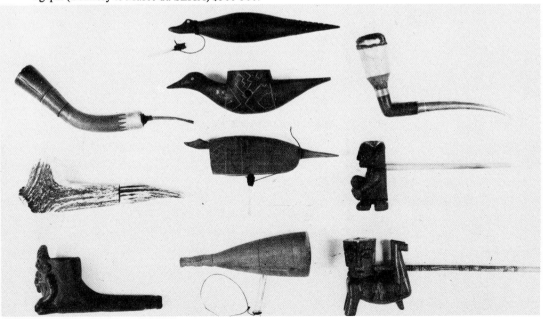

Oddities from around the world. Left, top to bottom: Montagnard wood pipe bound in copper wire, from Vietnam; antler pipe from Argentina; pottery from Guatemala. Center row: four wood effigy pipes from the Shipibo Indians, Peru. Right, from top: Montagnard pipe, brass and bamboo; two effigy wood pipes from Taiwan. (Author's collection) $100-200.

Tobacco smoking throughout the ages has inspired many inventors to dream up ideas to 'simplify' the process. Here is but a small selection of the thousands of proposals which could be classed as tobacco oriented. They may be difficult to buy in a local tobacconist's!

3362. Strauss, A. Oct. 1. 1874

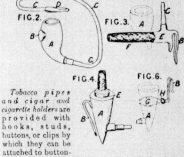

Tobacco pipes and cigar and cigarette holders are provided with hooks, studs, buttons, or clips by which they can be attached to buttonholes &c. Fig. 2 shows a pipe, the bowl A of which is provided with a button B. The pipe has a caoutchouc stem C and a mouthpiece D. A tube E, Fig. 3, which can be attached to a buttonhole by a hook B, is arranged to receive either a cigar F or a pipe bowl A. Fig. 4 shows a pipe provided with a hook B. A cigar holder E can be fitted into the pipe bowl. A cigar or cigarette holder may also be provided with a hook B and with a flexible tube and mouthpiece for smoking. A detachable clip has a hook B, Fig. 6, and arms G which are caused to grip the pipe bowl by a sliding ring H. The pipe bowl may have a projection which is passed through a buttonhole and secured by passing a pin through the end of the projection.

142,002. Crimmins, J. Aug. 25, 1919.

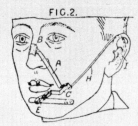

Tobacco pipes and cigar and cigarette holders. —A spring clip E, for supporting a cigar, cigarette, or pipe opposite to the lips of the smoker, is pivoted to a bent arm C, which is pivotally connected to a bent arm A, provided with a nose clip B, and to a stay H with an ear loop I.

18,915. Bergdolt, L. F. Aug. 23. 1911

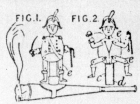

Tobacco pipes and cigar holders. — Upon the stem of a tobacco pipe, cigar-holder, or other suction tube is mounted a jointed figure the limbs e of which are connected to a spring piston d working in a cylinder c connected with the bore of the pipe or tube, so that, on reduction of pressure, the limbs of the figure move.

7666. Grabosch, G. April 25. 1900

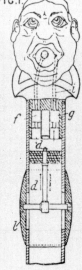

Cigar and cigarette holders. —Apparatus is provided in the handles of umbrellas, walking-sticks, and similar articles for keeping cigars or cigarettes alight. In Fig. 1 is shown a handle in the form of a human head with an open mouth capable of holding a cigar. By means of the piston-pump d operated by the ring l smoke is exhausted through the cigar and the valve f and discharged through the valve g and the passage i which emerges from the corner of the mouth. Instead of a piston-pump, bellows, vibrating diaphragms, balls, or the like, may be used.

1686. Parker, H. June 8. 1867

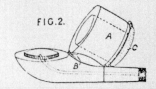

Tobacco pipes.—The bowl A is hinged to the stem at B. Thus the pipe can be recharged from the bottom without relighting, the lighted tobacco being pushed up to the top. The bowl is provided with a cap C.

Nineteenth and Twentieth century pipe dreams reproduced in Imperial Tobacco Group Review, Vol. 4, No. 2, February 1973. (Courtesy Imperial Tobacco Group Limited, England)

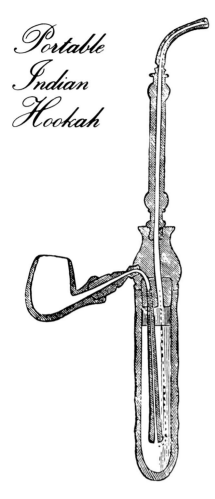

Portable Indian Hookah

ABOVE Several "new" pipes were introduced in 1870 and Skinners, a pipe manufactory in the Strand, London produced this portable Indian Hookah. One drawback of the pipe was said to be the continual gurgling noise when in use.

BELOW The Cottam Screw-pipe was another invention of the period. By adjusting the screw at the base of the bowl the tobacco plug could be moved upwards, thereby ensuring a smoke right through to the end and at the same time allowing any moisture to filter through to the base.

Cottam Screw-pipe

Two pipe inventions from the 19th century which appeared in Imperial Tobacco Group Review, Vol. 2, No. 3, May 1972. (Courtesy Imperial Tobacco Group Limited, England)

The realm of gimmick pipes is as infinite as the materials from which they have been made. A visit to a patent office would reveal some of the madcap and zany schemes for pipes I classify as gimmicks. . . Rube Goldbergian contraptions which, if used, would have drawn more attention than smoke! I cannot trace their successes or demises, but aware that they have their following, I have reproduced two diagrams I found in my research on the limitless designs for the pipe.

This chapter has briefly reflected on the many sometimes strange, sometimes comical configurations the pipe has assumed in various lands, among various peoples. There should not now be any doubt that the pipe represents a fascinating collector world, an empire with the broadest expanse of smoking implements from antler horn to emu's foot, from tusk to human bone, an empire over which the sun has yet to set!

FOOTNOTES: CHAPTER 8

102. Anthony Chute, Tabaco (Adam Flip, London, 1595) p. 16.
103. Ivor Noel Hume, A Guide To Artifacts of Colonial America (Alfred A. Knopf, New York, 1970) p. 308.
104. An Old Smoker, *op. cit.*, p. 30.

Chapter 9

YESTERDAY'S INVETERATE COLLECTOR–ART, INVESTMENT OR BOTH?

"The things, we know, are neither rich nor rare;
The wonder's how the devil they got there!"

Alexander Pope

More than ever, Americans are now turning to all sorts of antiques and collectables, buying them for investment as well as aesthetic appeal. One investment counselor cautions: "Collecting for profit is not for the timid or the uninformed, and the only way to be sure of winning at it is to buy things you can treasure personally even if they fail to make money."[105] But, the record has shown consistently that collectables, as investables, are beating inflation!

Pipe collecting is not a phenomenon of today. What is noteworthy are the personages who collected them. Duc de Richelieu, Armand Emmanuel du Plesis, 1766–1822, a French statesman who had served as Governor of Odessa, Russia and Crimea, later Chief Minister for Louis XVIII, had collected pipes. When he died, the value of pipes in the estate was appraised at cent dix mille francs; I suspect that was a large sum of money in those days![106] Another notable of the 19th Century, the R. H. Duke of Sussex, Prince Augustus Frederick, 1773–1843, sixth son of George III, had an immense collection of pipes. In June 1843, Mssrs. Christie and Manson of London auctioneered the sale of his estate which lasted seventeen days. The prices realized for snuff boxes and bonbonnieres were £2,238.12.6 and his 220 pipes, tobacco and cigars, £3,617.9.6.[107] Andrew Jackson, seventh President of the United States, was a noted pipe collector and many of Old Hickory's own pipes can be seen at the Hermitage near Nashville, Tennessee.

The arch collector of the late Nineteenth Century was a native of Birmingham, England, William Bragge, F.S.A., who became a master cutler in Shirle Hill, Sheffield, England. In one brief but energetic period of twenty years, he amassed over 6,000 pipes and tobacco accessories. His renown stemmed from exhibitions at Midland Institute Museum in 1870 and Alexandra Palace in 1881. Bragge died in 1884 but the collection was sold in April 1882 by Sotheby, Wilkinson and Hodge, and was thereby dispersed–a great portion went to the British Museum and the rest went into private collections. Today, an occasional piece reappears for sale and commands an extraordinary price. The very colorful commentary entitled "The Pipes of All Peoples" which appeared in the *Birmingham Daily Post*, December 16, 1870, is quoted for reader enjoyment:

> Some hundreds of people gazed with wonder on the magnificent collection of pipes of all nations, collected during many years by Mr. Wm. Bragge, F.S.A., the Master Cutler of Sheffield, and most kindly lent to the most important institute of his native town. From the wilds of Russia, the forests of South America, the dykes of Holland, the prairies of America, the valleys

of Germany and France, the coasts of Africa, Mr. Bragge has for twenty years collected every possible pipe. Whenever a traveller returns, like Catlin, or Petherick, or Burton, or Speke, or Grant, he is cross-examined at once on pipes, and all he can or will spare are secured, to be added to the vast collection of more than six thousand pipes—representing every age and every people in the world.

A sketch—even a sketch—of the contents of this marvellous museum of human tastes and industry and art, open this week only, may not only amuse but interest our readers, as well as those who have the greater pleasure of examining the curiosities themselves. Only a sketch at most can be given, for anything like a full description, in common justice to the extraordinary collection, would require a volume of very handsome size.

The early English pipes, extending in date from 1600 to 1750, have been collected from Dr. Thursfield's and other stores for many years. They include the rare varieties which have names and dates and marks, small in bowl, as tobacco was rare and dear, and increasing in size to a comely 'gauntlet' pipe, as the divine weed became more commonly known. Compared with these scores of examples, the early Dutch pipes contest for priority, and the Pipemakers' Guilds of Holland, and the private researches of Mr. Bragge, have collected an extraordinary mass. Here nearly all the stems are broken, but there are three early Dutch pipes, perfect and complete, so rare that no others are known; and while one of these was purchased, one was given and one bequeathed. Here, too, are elaborately-carved pipe cases, in wood and ivory, inlaid with metal, coloured, and gilt, in ivory, and lac, to contain the pipes, which once a year only, were taken by the Dutch traders into jealous Japan. Here, too, are early English snuff-mills, from 1607 downwards, when snuff was ground by private hands. Here are tobacco-stoppers, in quaintest forms and richest ornament of steel and silver, and some made out of Spanish dollars, too. Here is man's first rude mode of procuring fire, the long hard drill, turned quickly by clever hands on a softer piece of wood, till the lower piece was charred and burned, and fire was gained. Here are nearly every variety of fire-making tools, down to the pistol tinder boxes, so famous in Birmingham a century ago, with choice examples of Spanish and Dutch and Russian make. Here are tongs, and knives, and strikers, for pocket use—all connected with the use of tobacco during the last two hundred years.

The procelain pipes of Sevres, and Saxe, and Berlin, and Capo di Monte, and Furstenburg, and Copenhagen, and a score of other makes, are rich in every variety of decoration and every style of artistic skill. The English pottery pipes are wonderful also—long, tortuous, snake-like pipes, 'loudly' coloured and highly glazed, from Worcester, a century ago—one with a bowl for each day of the week; and dingy old brown glazed pipes of Brampton ware, with a dozen choicest specimens of the delicate white on a rich blue ground, with which Wedgwood has associated his name for ever. These by scores make up a most interesting collection of pipoceramic work.

Tobacco pouches of all ages and places, from beaded samples of American-Indian, to rough leather of Northern Europe, and exquisite lac and

A drawing of William Bragge with a montage of his pipes. (Courtesy of Pipe Line)

japan, of the East; from gorgeous hues of Syria and Persia, to simpler forms, of Iceland and Greenland, have been crammed into a case where brilliancy of colour rather than richness of style can be seen.

The European examples include, of course, every known form of pipe—Austrian and Styrian, in rich, grotesque, and quaint fantastic forms of unbarked stems; Italian pipes of delicate ivory, exquisitely and elaborately carved, and of steel, charmingly inlaid and damascened; German pipes of agates and similar choice stones, with meerschaums by the score; Finland pipes, embroidered with brilliant beads; Swedish pipes, of iron from the Dannemora mines; a fine Swedish Rune pipe; Italian pipes of the choicest Venetian glass; French pipes of porcelain and terra cotta and clay; and Roman pipes from the Campagna, in every conceivable variety of fantastic form. Asia contributes munificently in quality as well as style, to the great array. From the small metal-bowl pipes of China and Japan, richly decorated with lac and inlaying—sometimes of frosted silver, sometimes of filigree work, sometimes inlaid with silver, sometimes adorned with choice agates or rich crystal; from the pipe cases of Japan with their curious leather bags and belts, and richly ornamented stems of pipes; from the hardwood, silver-inlaid, amber mouthed pipes from the Caucasus, and Circassia, and Chinese Tartary; from the rich pipes of Persia, glorious in inlaying and brilliant in colour; from the superb perforated steel from Damascus, with opium-smoker's boxes from China, &c., down to the minor accessories of the smoking art, the collection is simply unrivalled, and full of interest to all. The choicest treasures of Asiatic art are found in the snuff bottles of China, where not only is all the mechanic skill of that patient people lavished on coral-coloured lac, and brilliant china, and splendid glass and porcelain, but magnificent samples of agates, and rock crystals, and chalcedony, and avanturine, and lapis lazuli, and jade of all colours, have been worked into fantastic forms, by untiring patience and consummate skill. Among the Japanese specimens are nearly a hundred 'nutchkies,' or buttons, used to fasten the tobacco pouch to the belt, and these, in finely-carved ivory, are worth an hour's study. Minute as much of the carving is, it is full of life and humour; full of pictures of Japanese life, full of real fun, and most characteristic of the real art of the Japanese. In fact, if this collection of Chinese and Japanese pipes, and pouches, and snuff bottles, and enamels, and glass, and porcelain, and silver, and metal work, and choice stones alone existed, it would be found a most valuable and comprehensive record of oriental art.

The Persian metal work—the hookah bases, to contain the water for the narghileh or hubble-bubble pipes—are simply the most magnificent examples of enamel work ever seen in our town, or procurable elsewhere. In beauty of design, and richness of colour, and perfection of workmanship, and brilliance of effect, these 'bases', and 'bowls', and 'stems' are beyond all praise, and offer an excellent example for rivalry, to modern Western metal-work art. The Eastern Archipelago presents examples not much less choice in the beautiful carvings of wood and ornamentation of brass and silver, from Sumatra and Borneo; while quaint pipes come from Jarva—one unbarked

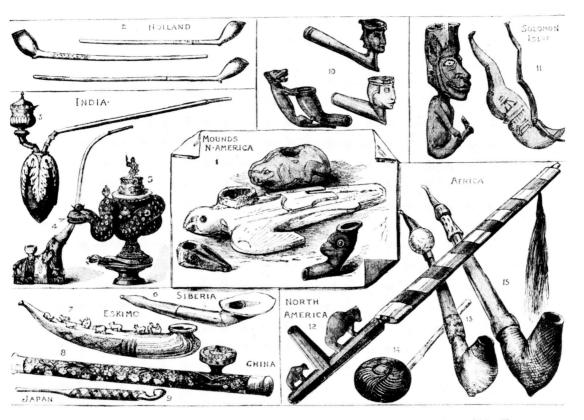

1. Stone and clay tobacco-pipes, the most ancient in the world (pre-historic) from Ohio, Tennessee, Virginia, and Pennsylvania.
2. Earliest European tobacco-pipes, Dutch, Sixteenth Century.
3. Indian, with silver-gilt fittings, the bowl, a nut.
4. Indian, the elephant of ebony, with gilt trappings.
5. Indian, the stand of green jade, the serpent of wood, silver fittings.
6. Siberian pipe, carved of mammoth ivory.
7. Eskimo pipe, carved of tooth of a cetacean.
8. Chinese opium-pipe.
9. Japanese pipe.
10. North American Indian pipes.
11. Betel box and cutter, from the Solomon Islands.
12. Calumet of Pawnee Indians, North America; the bowl carved of dark red stone, in the shapes of a bison cow and calf; the stem of painted wood, horsehair plume.
13. and 15. Pipes from the Nile, Upper Egypt.
14. Kaffir snuff-bottle, carved of hard wood.

(Courtesy of Pipe Line)

grotesque, one pipe belonging to William the Silent, having been presented with great ceremony to Mr. Bragge himself.

Africa also contributes a great and most curious collection to this great museum; from the wood and brass pipes of Algeria to the great gourd-pipes from Loando, and the huge mouth-pieces of the Bechuana pipes, to fit great thick lips; the thick fringe-decorated leathern-looking pipes from Ashantee and Dahomey, and Soudan and Assouan, and the White Nile (collected by Counsul Petherick); 'Fan' pipes brought by Du Chaillu; the wood and steatite pipes from Caffraria; the bamboo pipes, for smoking hemp, from Central Africa; the huge pipes from the Dutch Transvaal territory, of every size and shape, and every variety of barbaric decoration, form a most interesting attraction; such a collection never having been made before.

America—North and South—the original habitat of smokers, shows still more strikingly, but our space forbids the full details which the marvellous examples deserve. From the nameless 'grave mounds' of Ohio (collected by Squier and Davis), we have a small, hard, egg-like stone bored for the bowl, and perforated for the reed of a pipe; a dozen broken pieces of earliest pottery, wherein primitive tobacco was smoked; small fragments of forgotten races, dug out of mounds in Mexico and Central America; with dozens of superb specimens of the rich red stone of the great Pipe-stone Quarry, which George Catlin first visited and described; a sample of unfinished bowl from the quarry itself; samples inlaid in quaint style by patient hands; some oddly but naturally, carved with beavers and birds and sharks; some richly-broidered pipe-cases, of beads and hemp, and one with the bright, black, plaited hair from a woman's scalp; tomahawks, to hack the enemy and smoke the pipe of peace with; 'calumets' of every style, all full of 'character'; all revealing, not only the taste for tobacco, but the love of art, however rude in form, and giving, in fact, the true life of the Red Indian tribes.

From the north to the south of the two Americas, similar examples are shown; from the pipes made of mammoth tusk, among the Samoeds, where only driftwood is found, and where bone only will stand fire, and from the whalebone pipes from Sitka, and Greenland, down to the medicine-pipes of Paraguay, and the tinder-boxes made of the tooth of a tiger and the tail of an armadillo, every variety of American pipe is shown. From Vancouver Island, some most original pipes have come—flat dark slaty stone, carved in Mexican style, and some of wood, and some of bone also, while Sitka sends two quaint Noah's-ark-like-looking pipes, with a rude house on each, the chimney of which forms the bowl of the pipe, the stem the keel of the boat, and Noah himself seems to hold the helm, in most grotesque style—all made from bones of whales. Central America furnishes strange old pipes— many were in the collection of the unhappy Emperor Maximilan—and curously illustrating Mexican life and art. Far south, too, in the South Pacific, the New Hebrides group forwards from Solomon's Island some remarkable bamboo pipes and boxes, ornamented in curious style, in which smoking and chunam and betal chewing amuse the barbarous natives of those distant isles.

One large class of articles shown must be passed over with only a word of praise. The hundreds of snuff boxes in every variety of material and style will interest all. They include iron and steel, and metals inlaid with gold and silver, pearl boxes, damascened work, Roman and Florentine mosaic, agate, jade, onyx, fossil wood, birch bark, porphyry, enamel, lava, repoussé, relievo, &c.; while the ancient snuffgraters—the rappes of the seventeenth and eighteenth centuries—are marvellous examples of splendid carvings in ivory and wood, when snuff was ground by the taker, and was scarce and dear.

Such a collection, in a town like Birmingham, or any other town engaged in ornamental manufactures in metal, may be viewed apart altogether from the purposes for which the various objects are used, and as illustrating every style of ornament in use by savage or civilised peoples, as representing processes of every kind used to produce ornamentation, from the simple process of incision in wood, to the cutting of cold steel with chisel or graver, into designs grotesque or curious, elaborate and artistic. Many examples show how well their producers understood metalline sculpture, or sculpture in metal; of 'damascening', i.e., the insertion of tender threads of precious metals into other metals of lesser value, and by the art-labour employed therein, increasing the value of the object so enriched. There are many examples of great value and beauty, in 'niello'; and the enamelled examples give abundant illustrations of all its specialties of opaque and translucent, as applied on cloisonné or champlévé foundations; while the introduction of the translucent variety applied over the richly repoussé, or beaten-up ground, should have attracted the attention of all engaged in the production of articles of bijouterie, for personal decoration, in precious metal.

The Bragge Museum—as Mr. Bragge's generous loan well deserves to be called—includes more than six thousand specimens of art and industry from every quarter of the world, and of nearly every date. All relate directly to the use of tobacco, or some similar narcotic, and tend to show that 'the habit' is practically universal, and that the immortal memory of Sir Walter Raleigh is likely to be honoured when King James and his 'Counter-blast' are utterly forgotten and unknown.[108]

So impressive and variegated was the collection that a poem with the same title as the *Birmingham Daily Post* account appeared in *Punch* to coincide with the auction.

Baron de Watteville, 1824–1901, frequently referred to as the William Bragge of France, Director of Letters and Sciences to the Minister of Public Education and Commissioner of the French Expositions of 1867 and 1878, collected tobacciana and bequeathed his collection in 1912 to the Historical Museum in Berne, Switzerland. It is extant and intact today.

There was at least one of the banking magnate Rothschilds who collected pipes. Baroness Alice's collection, donated to the Museum of Grasse, France in 1927, is considered one of the most exquisite arrays of French faïence, porcelain, horn and ebony in Europe. Casual comments on collectors have always recognized Otto von Bismarck, not only as a heavy smoker but as a keen pipe collector reputed to have owned more than 300 exquisitely carved meerschaums.

At about the same period in the United States, a young Chicagoan, John F. H. Heide, began collecting pipes. In 1900, John Heide caught the collecting bug in

Munich and spent the next forty odd years travelling and corresponding with agents to obtain the finest specimens to represent a global expression in pipes. At his death in November 1946, the Heide catalog accounted for 1,344 unique specimens of the smoker's repertoire. George Revilo Carter of Chicago placed the following advertisement in *Hobbies* Magazine, October 1946: "To Museums, Ethnologists, Anthroponomists, Botanists, and their Friends. The Heide Collection of Pipes and other Tobacco Using Implements and Accessories is offered for sale as a whole, at inventoried cost. It will probably prove the last comprehensive collection to be liquidated. Assembled during the past forty-seven years, such a collection cannot hereafter be duplicated." It was said that the Heide Collection, at an inventory cost of $30,000, included at least one 4,000 year old pipe found near Boonville, Missouri; that was 2,500 years before the first mound pipes were discovered.[109] It comprised rarities for smoking opium, hashish and betel nut, American Indian pipes from the Queen Charlotte Islands to Florida, specimens from Manchuria, Siam, Japan, Korea, Indochina and a library of supportive literature—about 1,300 volumes on tobacco and smoking.

In December 1948, *Pipe Lovers* Magazine stated that Mr. Byron Cornell, a non-smoker, bought the Heide Collection of 5,000 pipes for $5,000. The final prices realized from the sale are not recorded anywhere to the best of my knowledge; what is certain is that Mr. Heide did not own 5,000 pipes and, more erroneously, the article states: "The pipes . . . were the remaining pieces from the famous collection of the late John F. H. Heide."[110] If 5,000 were the remaining pieces, then Heide had to have had more than 5,000—thus, the article is highly questionable! The Heide sale was the second in that city, preceded by an auction of high quality antique pipes owned by George Ellis Gary, a former vice president of the Brown and Williamson Tobacco Corporation. Following his death, his wife chose to dispose of it at auction on 25–26 February 1947.

In this century, whether for art, investment or both, other celebrities have owned extensive pipe collections: King Farouk I of Egypt, Edward G. Robinson, Bing Crosby, orchestra leader Lee Myles, comedian Ken Murray, Walter Woolfe King, star of "May Wine," and author Sidney Kingsley. It is rumored that the voice of Bugs Bunny, Mel Blanc, is a pipe collector. And the list goes on.

Precedent has been set and times have changed! Parke-Bernet 84, New York City, offered an outstanding collection of pipes and related material in a public auction on April 14, 1978. The catalog listed 292 lots, among which were 284 items from the Heide Collection, back in circulation. The 284 specimens were identified in the Parke-Bernet catalog, collated with the Heide list which is still available in facsimile; true proof were the stylized typed numbers which Mr. Heide had applied to each accession. A random sample of the prices realized in this auction, thirty years after the collection was dispersed, indicates that the greatest number of Heide rarities have appreciated considerably. The audience comprised not only collectors, but dealers, appraisers, scouts, inquisitive onlookers and art lovers. The auction realized some $55,000 with one meerschaum masterpiece carved by Arthur Schneider, Leipzig, reaching an impressive $3,400.

There are myriad pipes today valued at much more than $3,400. The noted Columbus Coming to the New World, a 32" aristocrat meerschaum depicting the

It is said that King Farouk I of Egypt had an immense collection of erotica. Obtained from that collection in 1964 is this turn of the century majestic meerschaum, more naïve than erotic! Eight nubile maidens, in high relief, adorn this pipe which measures 19½" long, 8" high. A close-up reveals the precise detail of this pastoral and picturesque scene. (Herbert G. Ratner Jr. Collection; photograph by Terry DeGlau) $2000-5000.

Italian navigator, six of his crew, a priest and Indians in full and bas-relief, carved by the William Demuth Company for the Columbian Exposition at Chicago in 1893, is now ensconced in the Valentine Museum, Richmond, Virginia. In 1936, that pipe was valued at $50,000.[111] And, the skilled hands of Gustave Fischer Sr. of Boston created the Battle of Bunker Hill, an immense pipe, 34" long with 31 figures carved in full and bas-relief, cut from a single piece of meerschaum weighing two pounds . . . this pipe was valued at $40,000 in 1906 and may appear, one day, in auction or at an estate sale![112]

Impressed with the overwhelming success of the April 1978 sale, the management of Parke-Bernet resolved to schedule frequent pipe auctions. There was, indeed, a repeat performance on May 22-23, 1979. Session One of Sale 690, the Collectors' Carrousel, included 45 lots of par excellence meerschaum pipes. I recommend a catalog subscription to all who desire to keep their finger on the pulse of future PB 84 specialty sales.

Excellent quality rare pipes, as any fine antique, are scarce and in increasingly greater demand. As the circle of pipe antiquarians enlarges, so does competition and with it, the spiralling effect on price. Prices have never been so high and there is no indication that they have crested.

"There is hardly anything in the world that some men cannot make a little worse and sell a little cheaper. And those who consider price only are this man's lawful prey."

                    John Ruskin

Gustave Fischer Jr. caressing the pipe his father took four years to make. The Battle of Bunker Hill, as this pipe is called, is based on an oil painting by the Connecticut artist, Amos Doolittle. However, Gustave carved the pipe using only a poor quality paper reproduction of that oil. Careful inspection reveals British troops, Colonists, the flags of both nations...three American and one British! The amber ferruled and turned stem has an American eagle surmounted. (Courtesy Mrs. Anna Fischer) $10,000-30,000.

FOOTNOTES: CHAPTER 9

105. Frederick Allen, "Better than Blue Chips: Objects to Appreciate." *New York Magazine*, Vol. 12, No. 9, February 26, 1979, p. 80.
106. Ana-Gramme Blismon, Manuel Historique et Anecdotique du Fumeur et du Priseur (Delarue, Paris, n.d.) p. 225.
107. Forrester, *op. cit.*, pp. 16-20.
108. *Pipe Line* Quarterly, Spring 1972, No. 11, p. 6; Summer 1972, No. 12, p. 8.
109. H. C. Hale, Jr., "The Heide Collection," *Pipe Lovers*, June 1948, pp. 176-178.
110. "Buys Pipes He Won't Smoke," *Pipe Lovers*, December 1948, p. 370.
111. *Long Island Daily Press, op. cit.*
112. *Boston Herald*, December 21, 1906.

Chapter 10

THE PUBLIC AND PRIVATE PIPE

> "To look upon a fine collection
> of pipes from different parts of
> our globe leads to meditation upon
> the mysterious attachment that has
> sprung up between mankind and a
> mildly narcotic plant."
>
>   A. E. Hamilton
>   This Smoking World, 1927

Perhaps in my lifetime there will never be an antique pipe collector's club or association. Without this vehicle, the tyro, lacking experience, counsel and the opportunity for information exchange, may falter and make costly financial mistakes. Without such an organization, even the mature collector will be deprived of a mart for buying, selling and trading.

This chapter provides sufficient guidance so that both the neophyte and the advanced collector can look before they leap. In the following pages I account for all the known permanent public exhibitions and on call pipe collections worthy of a visit. I also recognize a few private collectors in the United States with whom I have ongoing correspondence . . . many have collaborated with me and desire to share their knowledge with others.

Let us begin by a cross country trip through the United Sates; then, travel to Canada, Mexico and Europe. I shall describe the holdings in an unemotional, objective manner and in sufficient detail so that the reader can better decide where to go and what to see.

UNITED STATES

In New England, until late 1978, the jointly owned retail tobacco shops of Leavitt and Pierce, Harvard Square, Cambridge, and David P. Ehrlich, Boston, both of Massachusetts, possessed a panoply of tobacco related accessories. The former emphasized tins, labels, jars, rasps and political memorabilia; the latter exhibited well over 300 handsome meerschaums carved by Gustave Fischer Sr. at the turn of this century. The United States Tobacco Company, Greenwich, Connecticut, has since purchased both collections; by doing so, it has become the foremost industrially sponsored pipe and tobacco accessory museum in the nation. In this shade tobacco producing state, United States Tobacco opened its museum in a small clapboard cottage adjacent to the corporate headquarters on June 1, 1977. The initial holdings of about 2,000 objects stemmed from a purchase in 1974 of one very large collection from Mr. J. Trevor Barton, a British collector and tobacco archivist and a smaller collection of Twentieth Century carved relief briars known as the Marxman Heirloom Collection of Mastercraft Pipes, Inc. Those

Bountiful porcelain pipes at the US Tobacco Company Museum.

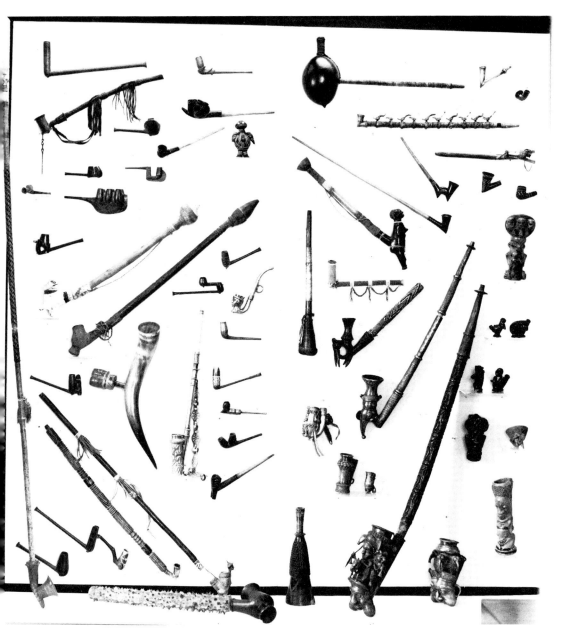

The fuming fantasies of Africa, US Tobacco Company Museum.

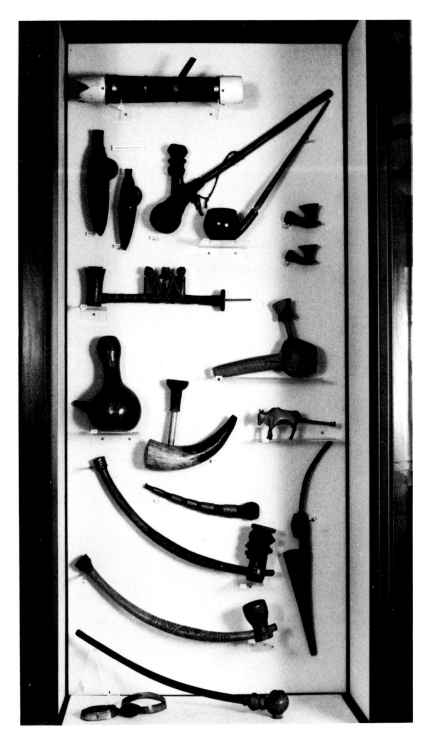

Another showcase of US Tobacco Company Museum Africans.

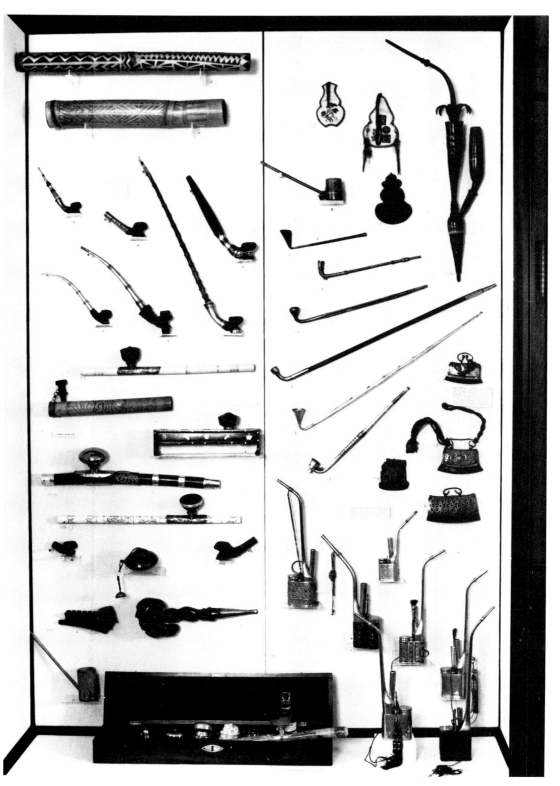

The Orient at the US Tobacco Company Museum.

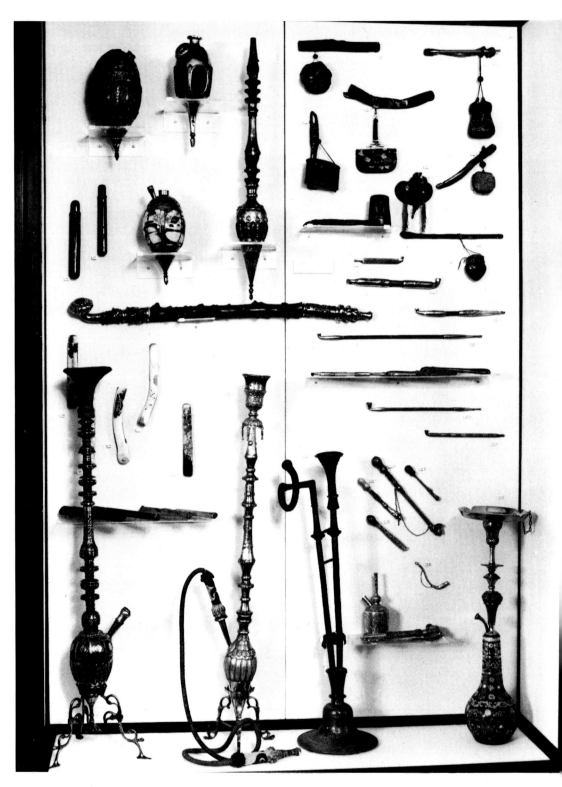

Nargilehs, netsuke tobacco pouches and other Near East smoking paraphernalia of the US Tobacco Company Museum.

acquisitions constituted the Borkum Riff International Pipe Collection and its value was then placed at $100,000. During 1976, prior to the museum opening, 226 representative pieces were selected for a travelling show and were displayed at shopping malls throughout the United States. With the 1978 purchase of the Leavitt and Pierce and D. P. Ehrlich collections, 1,139 additional items were introduced and the museum is now refurbishing the second floor of the cottage in order to ready these for public viewing. Ms. Jane Brennan, the curator, has been the dynamic force on a mission with purpose: to collect, perserve and herald the pipe so directly related to the tobacco industry. The museum is free to the public and open from 12:00 to 5:00 P.M., Tuesday through Sunday, except holidays.

Another museum now being established to demonstrate the historical importance of tobacco in the middle of the Connecticut Valley is the King Museum, Raymond Library Company, East Hartford. By the end of 1979, it is hoped that the currently small collection of material, principally cigar box labels and cigar bands, will blossom into an established and creditable collection of antique smoking pipes. Hours have not yet been set, according to the director, Mr. Ralph C. Secord, but it is intended that the museum be open to the public.

In New York, Indian pipes are conserved in the Cayuga Museum of History and Art, Auburn, the Brooklyn Museum of Art's Primitive Art Department, the Heye Foundation, Museum of the American Indian and the American Museum of Natural History. European pipes, once in the Metropolitan Museum of Art according to a 1958 *Spinning Wheel* article, are now not a prominent part of the museum's holdings and are probably in storage awaiting call up at special request.

In Pennsylvania, the Farm Museum of Landis Valley, Lancaster, has compressed two centuries, the Eighteenth and Nineteenth, into some thirty buildings and exhibition areas. For a nominal admission fee, visitors can walk through the restored structures and inspect the relevant tobacciana which dates from the last half of the Nineteenth Century and contains numerous smoking pipes.

Washington, D.C. has the Smithsonian and the Smithsonian has the permanent "We the People" exhibit at the National Museum of History and Technology. Opening on June 4, 1975, this bicentennial attraction includes folk and decorative art and military and political memorabilia linked to the American eagle, the flag, the Liberty Bell, Uncle Sam and the Goddess of Liberty. In the display, there is ample tobacco, tradition and Uncle Sam symbology which, by definition, includes aged corncobs and early American crafted smoking pipes.

Virginia, another tobacco producing state, has two noteworthy collections. The Valentine Museum, the Museum of the Life and History of Richmond, is the resting place of both the Half-and-Half Pipe Collection, donated by the American Tobacco Company in 1957 and the Duckhardt Collection of tobacco advertising ephemera, manuscripts and documents on the tobacco industry. In the vitrined Half-and-Half Collection sit stoically the William Demuth Company meerschaum bust series of U.S. presidents, started in 1881 with a likeness of President Garfield and eventually included Washington to Hoover; the famous Columbus meerschaum mentioned in Chapter 9; and a handful of exquisitely executed figural briars. In Danville, the Virginia State Corporation Commission chartered the Tobacco-Textile Museum for scientific, literary, educational and historical activities to provide an attractive and meaningful presentation of historical data and artifacts of both Virginia industries. Conceived in 1972 in an area which accounted for some 90 to 100 individual tobacco companies, the official opening occurred on September 16, 1975. Now open Monday, Wednesday and Friday from 2:00 P.M. to 4:30 P.M., the tobacco exhibit contains harvesting tools, cigarette packs, cigarette lighters and machines; the 800 or more pipes on exhibit include a few belonging to famous personalities as former President Gerald Ford and former Prime Minister Harold Wilson.

A third tobacco producing state, North Carolina, has recently begun to demonstrate renewed public interest in its tobacco history. Old Salem Inc., Winston-Salem, a restored Moravian Congregation town, includes the Miksch Tobacco Shop, completed in 1771—the oldest tobacco shop still standing in America—restored and opened in 1960; it is noted not for its antique pipes as much as it is for its tobacco selling lineage. Although the pipes in this collection are judged as being just ordinary, Mrs. Martha Battle, Collections Assistant of the North Carolina Museum of History in Raleigh, discloses that the museum accessions include both antique meerschaum and silver pipes.

Most approximating the reader's interest is the Metromedia Collection of antique meerschaums in the Museum of Art on Duke University campus in Durham. The collection comprises 67 exotic and exquisite meerschaums once belonging to Major Max C. Fleischmann of Santa Barbara, California and Reno, Nevada, a world traveler, big game hunter and sportsman. Upon his death in 1953, it was bequeathed to Richard Schutte, Vice President of Metromedia Inc., New York City, from 1960 to 1970. The President of Metromedia, John Kluge, bought the collection and displayed it in the headquarters until it was donated to Duke in 1976. The collection is at center stage surrounded by pre-Columbian and early African ethnographica. This museum is open to the public from 9:00 A.M. to 5:00 P.M., Monday through Friday, 10:00 A.M. to 1:00 P.M. on Saturday and 2:00 P.M. to 5:00 P.M. on Sunday . . . well worth the trip! The Duke Homestead, Durham, one of North Carolina's historic sites, is the ancestral home of a family whose name is synonymous with the tobacco industry. Once property of Duke University in 1931 and presented to the State in 1973, it is now a tobacco museum and farm containing exhibits which should be seen as part of that state's tobacco culture. The last major display in North Carolina is a collection of Indian pipes belonging to the Eastern Band of Cherokee Indians at the Museum of the Cherokee Indian, Cherokee. This museum is open to the public.

In Florida, two museums attract daily visitors. The Lightner Museum in St. Augustine, representing the World of Yesterdays, has 17 Victorian shops on its first floor; one, the Tobacconist, contains a wealth of tobacco accessories, including a variety of old smoking pipes. The St. Petersburg Historical Society Museum proclaims an extensive collection of pipes in two vitrines—meerschaums, orientals, porcelains, clays, chibouques, Indian peace pipes and assorted woods.

In the heart of downtown Chicago, a nationally reputed retail establishment, Iwan Ries Company, recently celebrated its 122nd birthday in the Chicago Loop, having served the smoking public for that many years. In an alcove of the store is a museum for the antique pipe collector. Although not receiving the acclaim he rightfully deserves, Stanley B. Levi, Chairman, should be extolled for having assembled a breathtaking exposition of fine quality pipes, mounted in shadow boxes, in full and open view. With the use of a company furnished cassette recording, the visitor is taken on a self-paced, leisurely walk through the past replete with the splendor and beauty of the European meerschaum, the rustic and the crude, pipes amassed over the years by a family of five generations steeped in the retail tobacco tradition. This experience is not duplicated anywhere in the United States.

In Wisconsin, the Kinosha County Historical Society Museum acknowledges a quality collection of pipes and snuff bottles but no further details could be provided. Of more importance is the Milwaukee Public Museum. It preserves the famous George A. West collection of American Indian pipes (see Chapter 12) which can be accessed and studied by both the student of anthropology and the collector of Indian artifacts. The collection originally numbered over 1,000 and was considered one of the finest in the country at the turn of the century.

The Indian Shrine Association, Pipestone National Monument, Pipestone, Minnesota, occupies the site of the famed pipestone quarries where Indians obtain the famous red stone still reserved to them today by special law. Plains, elbow, hatchet, disk and micmac pipes can be seen, the old as well as those made daily for resale at the Cultural Center.

In Iowa, the Yesterheim Norwegian-American Museum, Decorah, acclaims a broad collection of European tobacciana dating from the Eighteenth Century: snuff, tobacco jars, cutting boards, snuff bottles and a respectable number of old clay and meerschaum pipes. For the historian, H. M. King Olav V of Norway is the Honorary Chairman.

In Colorado Springs, until recently, the James Lee Dick pipe collection of Hutchinson, Kansas, could be seen in the Pioneer's Museum, founded in 1908 in that Pike's Peak city. The Dick Collection numbered over 600 world-wide pieces with two-thirds on display. My 1979 correspondence to the Pioneer's was stamped returned to sender and prior contact with that museum was in 1976.

Lastly, in California, the writer is most familiar with a collection growing at a geometric rate—the Tinder Box International Limited Pipe Museum at the corporate headquarters, Santa Monica. At the helm is Doug Murphy, head of Retail Sales and Services who has vigorously pursued gathering a representative collection of species pipes for this nationwide chain of retail stores. In time, the TBI collection will soon compete with and may surpass the aforementioned collections if Doug Murphy is allowed enough money and the time to spend it!

CANADA

In Montreal, the visitor can spend the greater part of a day visiting McGill University's McCord Museum which has a spread of North American Indian pipes with the expected flavor of contributions from tribes living on the borders. Then, he should visit Château Dufresne, Musée des Arts Décoratifs de Montréal, which acclaims some 3,000 accessories donated by David Macdonald Stewart, a former President of Macdonald Tobacco of Canada. The collection was formed in the early Twentieth Century and the museum is now in the process of cataloging. There is a sister collection to the International Borkum Riff Collection, titled the Borkum Riff Collection, Imperial Tobacco Products Limited in Montreal. This collection was the precursor of the one in Greenwich, Connecticut; it has less depth and breadth but is an impressive collection of approximately 300 pieces, valued at $100,000 in 1973. In 1974-1975, the collection travelled throughout the provinces and now lies at rest at the corporate headquarters, housed in six walnut cabinets for easy viewing. The National Museum of Man in Ottawa is an archaeologist's dream with hundreds of specimen prehistoric pieces; sadly, they are not on public display. As well, the Provincial Museum of Alberta in Edmonston boasts a pipe collection of ethnographic specimens, hundreds of stone pipes used by the Province's Indians and some clay pipes used by fur traders.

MEXICO

Only one museum collection, at the moment, is traced to South of the Border. El Museo Nacional de Antropología, Mexico City, has a group of prehispanic and preclassical pipes in its Aztec Culture Room representing various territorial states.

EUROPE

For the pipe collector, Europe is the mecca of tobacco museums, a paradise which evokes collector awe, inspiration, admiration and envy. The nation, the state and the tobacco industry have managed to pool their combined efforts to enhance, not stifle, the cultural and social significance of tobacco and its utensils. I believe that once having read this section, the reader will be convinced.

ENGLAND

The home of the clay pipe enthusiast, anthropologist, archaeologist, ethnographer and researcher should be the very place where the greatest accumulation of clay pipes is found. In the British Isles, this is so!

The London Museum, Kensington Place, merged with the Guildhall Museum in 1975 and reopened in 1976 as the Museum of London, bringing together 2,000 years of London city history. In it are found the typological clays of London, archived by two clay authorities, David R. Atkinson and Adrian Oswald. The pipes range in date from 1580 to late Nineteenth Century and are available for inspection by the public. The British Museum is doubly interesting. Its Great Russell Street establishment has a good general pipe inventory, the Oldman and Christy Collections which represent a universal expression of the tobacco pipe. The Ethnography Department, Museum of Mankind, British Museum, Burlington Gardens, claims some 500 tobacco pipes from North America, stemming from two additional collections, that of William Bragge and one from E. G. Squier—E. H. Davis. Of the pipes in the Bragge Collection, some 160 come from North America and were purchased in 1882 for the Christy Collection along with an elaborate manuscript which Bragge commissioned George Catlin to produce around 1864-6 to show the forms and use of the pipe. The Squier-Davis Collection is principally of archaeological specimens from Ohio in the 1840's. The latter collection was housed in the Blackmore Museum, Salisbury, until 1931 when a large fraction was transferred to the British Museum and eventually to the Museum of Mankind. In all, the pipes range from 100 B.C. to 1900 A.D., from Ohio to the Northwest Territories—colorful, historical, educational and without question, priceless.

One industry museum in London deserves a place of honor. The Dunhill Museum, St. James's, is already internationally renowned. The family is responsible for authoring two important treatises: THE PIPE BOOK in 1924 and THE GENTLE ART OF SMOKING in 1954. Beyond that, the collection, amassed over a great number of years, comprises the fullest range of the pipe by age, style, material and value in the British Isles. One can never be sure he'll catch a glimpse of the entire collection at one viewing, since many of the pieces cross the Channel, with frequency, to other Dunhill enterprises on the continent. Yet, at whatever time the viewer spies the collection, he is confronted by a formidable collection.

Iwan Ries' counterpart in England is Astley's, Jermyn Street in London. Astley's is a retail establishment, selling pipes, cigars, tobacco and . . . antique pipes. One need only correspond, describe his want and price range and who knows? Astley's most recent acquisiton was an elaborately carved meerschaum dating from the 1870's, dubbed Pipe of Peace, 1870–1871, commemorating the treaty ending the Franco-Prussian War. In its elaboration, Queen Victoria, Gladstone, Napoleon III and other dignitaries can be identified. This, too, may be for sale!

The Barling Museum, Liverpool, at the Headquarters of B. Barling & Sons, Ltd., owns a less than formidable but quality exhibit of first-rate antique smoking pieces and accessories.

Bramber, near Steyning, Sussex, may be otherwise, an historic landmark, but it is certainly a haven among pipe aficionados. Established in 1861 as Potter's Victorian Natural Science Museum, it was reopened in 1973 as the House of Pipes, a smokiana exhibition. Anthony "Tony" Irving who has collected pipes for almost his entire adult life, put his labor of love on public display—his 20,000 plus collection valued at £250,000—he has seen his ambition, only a germ in 1948, come to fruition. To further the theme, Tony has included an executive suite, a conference hall, a library, study center and gift shop and modeled one room as a Seventeenth Century tobacconist's coffee shop. Tony's reputation is exceeded now only by the kudos and plaudits from the media and from the thousands of daily visitors to this outstanding exhibition.

Though the W. D. and H. O. Wills Tobacco Ltd., Bedministor, Bristol, prides itself in an extensive library of over 10,000 books, 10,000 pamphlets and 400 periodicals on the tobacco industry, the author's records include recognition of an extensive and choice, albeit low key, collection of 6,000 pipes of all ages and peoples. It is not known if the corporate display is available for public viewing or only by appointment. Carreras Rothmans Ltd. of Basildon, Essex, has amassed one of the largest collections of British tobacco ephemera: cigarette packs, labels, show-cards, posters, advertising memorabilia—and, pipes. Although not competing with Alfred Dunhill or Irving's House of Pipes, the archives contain sufficient selective and rare specimen pipes to draw attention when the curator assembles an historical display.

As an aside, for those collectors of treen, Tunbridge and Scottish woodware, the superb Edward H. Pinto collection can be seen at the City Museum and Art Gallery, Birmingham. Among the 6,000 to 7,000 objects are pipe racks, stands, trays, walking sticks, spittoons, tampers and a host of Eighteenth–Nineteenth

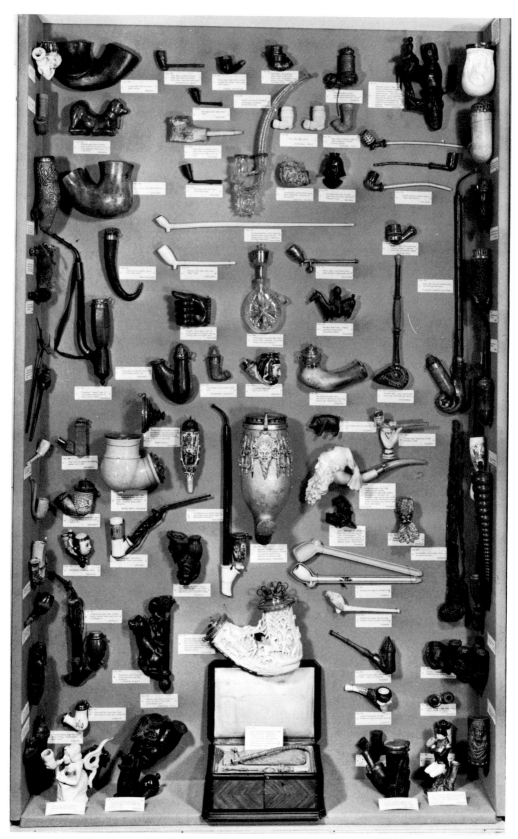

Six showcases of the Dunhill Pipe Collection, London. (Courtesy Alfred Donhill Ltd.)

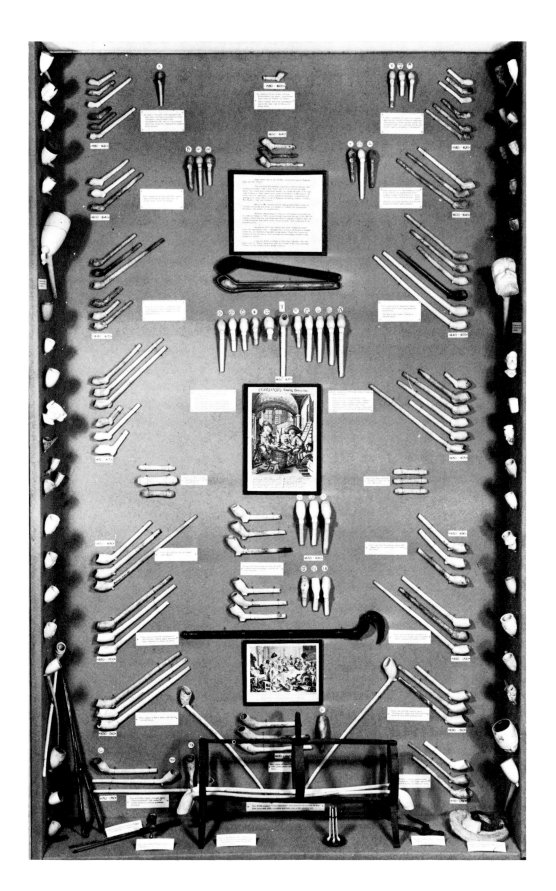

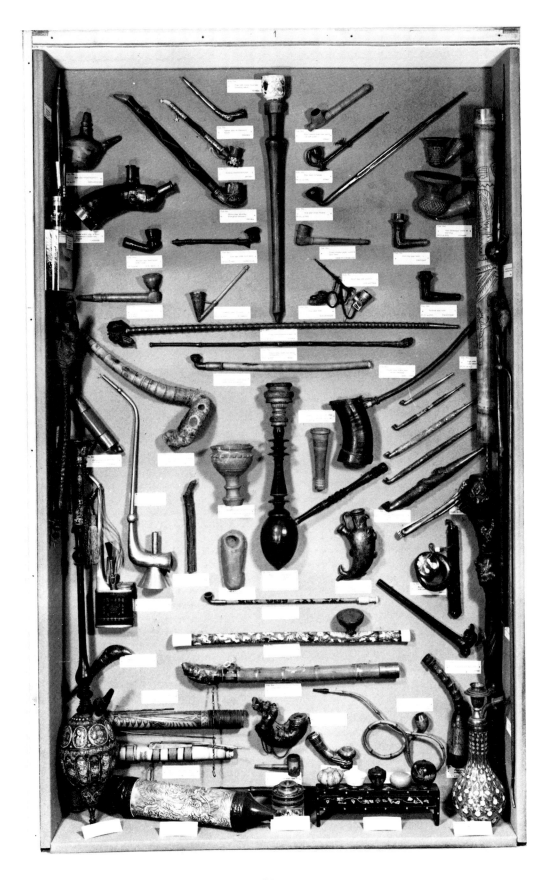

Century European hardwood pipes, pipe moulds and pocket cases for clays ... an outstanding assemblage of fine carved wood utensils for the smoker-collector. As well, there is a display of equipment belonging to two clay pipe makers of that city.

And, now by a very circuitous route, I arrive at the English clay pipe. Without elaboration, the number of museums which honor it is extensive, so my notes are brief only to whet the appetite of all would-be clay collectors: the Lincoln City and County Museum, Lincoln; Kingston-Upon-Thames Museum; University Museum of Archaeology and Ethnology, Cambridge; County and Folk Museum, Cambridge; High Street Museum and Christchurch Mansion, Ipswich; Pitt Rivers Museum, University of Oxford, Oxford; Museum, Castle Arch, Guildford; Thursfield Collection of Broseley Clay Pipes, Clive House Museum, Shrewsbury; Gorey Castle Museum, South Hill, Jersey, C.I.; Taunton Castle, County Museum, Somerset; Horniman Museum and Library, London; Wilberforce House, Kingston-Upon-Hull; Museum, Salisbury and South Wiltshire; Grantham Museum, Lincolnshire; the Castle Museum, Colchester and Essex; Norris Museum, St. Ives, Huntington; at Castle Museum, York, there is a reconstructed workshop of clay maker Sampson Strong; Abbey House Museum, Kirkstall, Leeds, containing Strong's claymaking equipment; Wisbech and Fenland Museum, Wisbech; Museum and Art Gallery, Petersborough; Cecil Higgins Art Gallery, Bedford; Museum and Art Gallery, King's Lynn; Castle Museum, Norwich; Boston Museum, Lincolnshire; Stewartry Museum, Kirkcudbright, Scotland; and, Huntly House, Cannongate, Edinburgh which has a reconstructed workshop of a local clay pipemaker, William Christie.

WESTERN EUROPE
FRANCE

Throughout the country, as one is accustomed on the entire continent, just about every museum has a token vitrine or showcase of essential tobacco artifacts. I list only the principal museums here. First, the museum which has received the most international acclaim is le Musée du S.E.I.T.A., the Service d'Exploitation Industrielle des Tabacs et des Allumettes, Paris. Housed originally in a galerie adjacent to the headquarters, a renovated museum was inaugurated on January 18, 1979 as the museum dedicated to the history of tobacco and its usages. S.E.I.T.A. has been, over the years, of global importance to tobacco, emphasizing the culture, the folk art and the traditions of tobacco use through the media, public exhibitions, feature stories in its quarterly magazine, *Flammes et Fumèes* and occasional promotional material. To inaugurate the renovation, its grand opening theme was "L'Affiche et le Fumeur," the Poster and the Smoker, which ran from January through March 1979. The scope and intensity of the S.E.I.T.A. efforts are unparalleled on the continent. In Paris, there is also le Musée des Arts et Traditions Populaires and a prominent gathering of snuff rasps, tobacco pots and pipes.

In Nice, le Musée Massena contains an interesting assortment of pipes offered by the Emperor. And in Rouen, le Musée de Ferronnerie, Le Secq des Tournelles, reputed as the single largest collection of forged ironware in Europe, conserves rare iron pipes and other iron tobacco utensils.

Le Musée de Grasse, Alpes-Maritimes, the donee of the Baroness Alice de Rothschild collection, had tallied 232 superfine pipes at last count but, regrettably, the collection is not available continuously. Le Musée de Saint-Omer, located in one of the key French tobacco centers of the Eighteenth Century and home of the Fiolet clay pipe of the Nineteenth Century, contains remnants of the famous clay pipe factory, Duméril Leurs, of that city. In Ville de Bergerac, Département de la Dordogne, tobacco capital of France, le Musée d'Intérêt National du Tabac was inaugurated on September 4, 1950 under the auspices of the French Direction of Museums. Four exhibition rooms, each creating the aura of tobacco culture, welcome the visitor: one consecrated to tobacco in history, politics and customs; a salon consecrated to the culture, manufacture and commerce of tobacco; another consecrated to the tobacco industry; and, the last to manufactured tobacco products. In each, alongside snuff, cigars and cigarettes is placed the exhalted pipe in every representative form.

Saint-Claude in the Jura Mountains is the home of the French briar pipe. Long the center of wood turners, the region is steeped in the history of pipe carving. La Confrérie des Maîtres-Pipiers de Saint-Claude is a guild of 19 local manufactories, an association, a brotherhood. Maître-Pipiers, master-piper, is a title awarded only to deserving manufacturers; Confrères-Pipiers are elected by the Confrérie for their individual contributions to pipe smoking and the industry... of the millions of pipe smokers in the world, there are only some 200 Confrères in its Hall of Fame. As it should be, the Confrérie has a museum, le Musée de la Pipe, devoted to erica arborea in all its glory.

SPAIN

One accredited retailer, José Pozito, with six establishments in Madrid, has been making valiant efforts to build a representative collection of pipes for public viewing. He first drew the industry's and my attention when he went on expo in late 1976 and then hosted a writer's competition in 1977, offering 250,000 Pesetas as first prize for the best literary account of the history and culture of the pipe. Pozito's is most active in spreading the word on the growing attachment of man to pipe. In the Museo del Pueblo Español, Madrid, and in the Museo Federico Marés, Barcelona, there are smokers' collectanea.

GERMANY

The Deutsches Tabak und Zigarrenmuseum in Buende, Westfalia, claims, among other things, a giant cigar manufactured in 1936, 1.70 meters long, weighing 18 pounds and estimated to burn for 600 hours even at the pace of the most avid puffer. The Focke Museum, Bremen, houses the expansive Martin Brinkmann AG tobacco collection; key to the famous Tabakskollegium of Friedrich Wilhelm I of Prussia, 1688–1740, is the luxurious assemblage of Meissen, Delft and wooden tobacco jars. In the small town of Tauberbischofsheim, the Heimat Museum has a permanently encased display of porcelain, regimental and bavarian pipes donated by a retired local construction worker. The Germanisches Nationalmuseum in Nuernberg frequently exhibits a small but respectable pipe collection by scheduled cycle. Haus Neuerburg, a tobacco firm in Cologne, has an immense collection of pipes, tobacco jars and snuff rasps, once the subject of a richly pictorial and rare catalog printed in 1930. And, H. F. and Ph. F. Reemtsma of Hamburg, another manufacturer has, over the years, amassed a world-wide tobacco treasure-trove of valuable pipes for public viewing.

THE NETHERLANDS

As in Germany and France, the Dutch have displayed an enthusiasm and affection for tobacco culture and history. The principal attractions are industrial and state museums and one private collector who went public to unfold his enormous clay pipe collection. The Niemeyer Company, manufacturers of Flying Dutchman, Sail and other tobaccos, had maintained a tobacco cottage-museum facing one of the countless canals in Amsterdam. In 1975, the museum was relocated to its headquarters in Groningen. Now, Niemeyer exposes the collectable pipe in three settings—a tobacco-shoppe, an apothecary and an inn—the pipe's normal habitat of yore. Following suit, Douwe-Egberts, the Amphora people, a firm specializing in tobacco, tea and coffee, now well over 200 years old, owns a museum, De Witte Os, The White Ox, to exhibit three centuries of bygones—tobacco introduced to Holland in the Sixteenth Century and tea and coffee, introduced in the Seventeenth Century. Under the aegis of its affable curator, A. Witteveen, the pipe room recreates an old-fashioned tobacco shop—Delft and Makkum tobacco and snuff pots, countertop tobacco figures and pipes of every shape and size. Admission is free to visitors with the proviso that they obey the "No Smoking" signs. The J. & A. C. van Rossem Company, makers of Troost tobacco, has a lesser known collection of antique tobacco artifacts which can be seen in Nijkerk, by appointment. Gouda, where 300 years ago clays were made, is famous today for its cheese markets and for a municipal museum, The Moriaan, Blackamoor. It contains an old tobacco shop with its original inventory of Delft tiles, Gouda Stoneware tobacco jars, the utensils of a typical Gouda clay pipe-maker and a representative collection of Goedewaagen clays.

Inside the White Ox Museum. (Courtesy Douwe Egberts' Pipe Room, Utrecht, Holland)

Another view of the White Ox. (Courtesy Douwe Egberts' Pipe Room, Utrecht, Holland)

The collector mentioned earlier, Don Düco, has been a long time collector of clays who, by my count, must now own more than 5,000 specimens from Holland, Germany, France, Belgium and England, the principal countries which fabricated the clay. On June 18, 1975, Don opened Pijpenkamer Icon, the Icon Clay Pipe Museum, at 16 Fredericksplein, Amsterdam. The timely opening of Pijpenkamer Icon filled the void in Amsterdam's tobacco culture left by the closure of Niemeyer's Museum in that year. Don's museum has received glowing reports in just about every tobacco periodical published today. Don is not the average collector; he is also a prolific writer, researcher and historian. He encourages visitors and correspondence from anyone around the world who can converse in his language—clay!

LUXEMBOURG

In Luxembourg, Maison Heintz van Landewyck, manufacturers of tobacco and cigarettes since 1847, owns a corporate museum with a wondrous array of pipes representing Old Europe.

Two views of De Moriaan Museum, Gouda. (Courtesy of the Museum)

BELGIUM

In Belgium, le Musée de la Vie Wallonne, Liège, traces the development of clay pipes in that country for two and one half centuries via a reconstituted atelier of a clay manufacturer, vitrines bulging with clay fragments from domestic factories, posters, advertisements and original moulds from Belgian pipe makers... pleasant environs indeed for the student of the clay pipe. And, in le Musée du Tabac et du Folklore, Vresse-sur-Semois, once again, many hundreds of pipes, tobacco jars and boxes, pipe cases and antique smoker lithographs surround the collector in an apropos setting.

SWITZERLAND

In Switzerland, the Berne Historical Museum, Château Oberhofen, was, as stated earlier, the legatee of the Baron Oscar de Watteville pipe and tobacco utensil collection in 1912. It came to my attention in a special showing exactly sixty years later, in 1972, with the unveiling of 23 vitrines representing the Baron's lifetime endeavor, paralleling his English rival, William Bragge.

In Geneva, two collections can be seen. Fondation Alfred and Eugénie Bauer-Duret and La Collection Rhein. Little can be said of the first, but Collection Rhein is, at present, a fixture in the show windows of Tabac Rhein, a retail establishment on Rue Mont-Blanc. A resounding bravo and heartfelt thanks need be expressed to Monsieur Jean-Charles Rhein for having amassed a collection of first-rate meerschaums, porcelains and woods. If the viewer cannot visit, he can purchase the Rhein catalog and admire the pipes from afar in a superlatively executed full color photographic guide (see Chapter 12).

AUSTRIA

In Austria, the pipe collector's world is just as bright and promising. In a small town near Salzburg, there is a beautiful and fashionable hotel on the peaceful and sunny peninsula of Lake Fuschl. Fishermen may enjoy secluded grounds; hunters may ensconce themselves in any of five hunting lodges; there is a golf course, horseback riding, mountaineering and gourmet eating for all. Moreover, in the farm loft of this Seventeenth Century Jagdgasthof, is a hunting museum. Its fine exhibits include world record hunting trophies, historical poachers' weapons, dioramas, hunting paraphernalia and one of the largest pipe collections in the world, some 1500 in all, keyed to the themes of hunt, wild game, field and stream—in porcelain, meerschaum, metal, wood and staghorn. So prestigious is the entire exhibit that a book was devoted to it in 1974—JAGDSCHAETZE IM SCHLOSS FUSCHL. One chapter is devoted to these prizes of smoking.

In Vienna, yesterday's center for the infinite expression in meerschaum, the Museum of Austria Tabakwerke A G, under the curatorship of Dr. Fritz Hinterndorfer, is a totally fascinating and awe inspiring world for the antique pipe collector. The magnificent collection is available to the public on corporate grounds which date back to 1903–1905. The firm claims that the collection began after 1873, the year of the Vienna World Exposition. From primitive to prissy, pipes of every manner and matter are arrayed alongside rare oils and watercolors which depict the smoker.

SCANDANAVIA

In Copenhagen, Ole W. Ø. Larsen, a fourth generation entrepreneur in charge of a 113 year old pipe manufacturing business, has his own museum on the premises to feature the Dane's own tradition of tobacco and smoking. Ole's father had put the collection together and Ole popularized it, enlarged it, added a tobacco library of first editions and enhanced it with paintings, etchings and advertising memorabilia from Scandanavia. Ole and his pipe smoking daughter Anne encourage visitors since, between them, they can master seven languages and get their message across about the history and provenance of each museum piece. Little is known of another popular pipe collection in Copenhagen, Storm P. Museum, but it is available for inspection. In Odense, Erik Stokkebye A/S Cigar and Tobaksfabrik, founded in 1882, owns a first-rate pipe collection, but at the moment, it is cached. Jaegerspris Palace, Jaegerspris, exhibits the grand collection of meerschaums from King Frederik VII of Denmark, an avid pipe smoker and collector.

Sweden's Svenska Tobaks AB, the Swedish Tobacco Company (STA), has renovated an old restaurant called Gubbhyllan, the Old Men's Retreat, in Stockholm's open air museum, Skansen. Gubbhyllan was an old country inn built around 1820 and became a well known restaurant and konditori. In 1965, STA moved the house to Skansen and converted it into a tobacco museum to display STA owned artifacts. On the ground floor are the selling, packaging and advertising ephemera; on the second floor are advertising signs, wooden Indians, a very large collection of pipes, snuff boxes, cigar cutters and a typical smoking room of an 1890's middleclass gentleman; in the attic are found tobacco curiosities, a tobacco factory office and a cigar maker's lodgings from the end of the Nineteenth Century; the basement, constructed in conjunction with the move to Skansen, is devoted to the history and cultivation of tobacco and the manufacture of tobacco products with an emphasis on Sweden. This fantasy collector haven is open all year round for Skansen is a modern recreation center visited by one million and a half visitors each year.

In the National Museum, Helsingfors, Finland, there are, conservatively, several hundred pipes and other smoking utensils. At Oy Strengberg AB Tobacco Corporation, Jakobstad (Pietarsaari), some ninety antique pipes can be seen, among them early clays from 1590, a gift from Fribourg and Treyer Tobacconists in London, England.

The National Museum of Iceland, Reykjavik, houses an assemblage of old clays and pipes of baleen and ivory in one of the Nordic exhibition rooms.

EASTERN EUROPE AND THE EAST

I have read of famous pipe collections in the National Museum of Czechoslovakia at Bratislava and the National Museum of Hungary at Budapest. The newest pipe dream is northeast of Budapest at the small village of d'Eger, where a fortress has become the home for more than 10,000 Sixteenth and Seventeenth Century Dutch, Turkish and Hungarian pipes. D'Eger is undoubtedly the most unique collection in Central Europe. There is also a respectable collection of pipes at the National Museum of Tokyo. Sadly, I cannot discourse on these last museums

The Tobacconist's Gallery Shop, ground floor, Gubbhyllan. Skansen, as it appeared at the turn of the twentieth century. (Courtesy Svenska Tobaks AB)

and I leave it to a global traveller to recount the smoking wonders found at these historic sites.

EPICUREANISM

Since the tavern, pub, inn and coffee-house of the past catered to the indulgent triumverate of food, drink and smoke, I add a parenthetical note to call attention to a few establishments known to patronize the pipe fancier. The Pipe Dream Restaurant in West Willington, Connecticut, according to its owner, Bert Barone, is a quaint restaurant with a rustic country atmosphere situated near the Massachusetts border. The main dining room is decorated with paintings in all mediums portraying smokers and smoking scenes and some 300 antique pipes—a delightful setting for the gourmand and the pipe epicure. Keen's English Chop House in New York City was known as the Lamb's Club at the turn of the century when it was a rendezvous for the theatrical crowd. The custom of following dinner with a pipeful has survived and I had the pleasure of eating there in 1972 as Pipe Register member 90207. For those who have never had the experience, I quote the membership card: "...is hereby identified as a life member of that illustrious company who follow the ancient custom of calling for their 'clay' after dinner at KEEN's. As such, he is to be accorded special consideration. His 'churchwarden,' bearing the above number, is in our custody and reserved for his use whenever he partakes of this tavern's hospitality." Today, there must be more than 100,000 members in the Register, their clays hanging from the rafters! On July 6, 1977, one of the renovated warehouses in Richmond, Virginia's Shockoe Slip on E. Clay Street became the Tobacco Company. The idea belonged to Jearald D. Cable, a former vice president of the Medical College of Virginia, who now is a managing partner of Richmond's hottest spot for drinking, dining, dancing and tobacco artifacts. The design features are reminiscent of old tobacco firms and warehouses and the renaissance in memorabilia and good food has brought three floors of excitement to that city.

Those going abroad can dine in the more traditional and rural English clime of cakes and ale at two taverns which have come to the attention of antique pipe lovers. Each has an extensive array of antique pipes to view, compliments of the respective owners; they are The Wise Man Pub in West Stafford, near Dorchester, Dorset and Pipe's Tavern in Newsham, Yorkshire.

THE PRIVATE COLLECTOR

The list below, although short, represents the more active and vocal American pipe collectors with whom I am acquainted. Only their names and specialties appear; details as biographical sketch, address and telephone number will be furnished only upon written request to me in care of the publisher. This procedure has been invoked to protect the interests of each collector. There are many more pipe collectors in the United States but they have expressed a desire for anonymity and I have respected that right:

Bill Caruth	C-O-Gs
Michael P. Clark	supersize meerschaums
Thomas E. Clasen	antique pipes; WDC articles
Ed Clift	old pipes, smokers' accessories, related bibliomania and ephemera
Martin M. Cohn	all classes
Doug Diez	clays
Ken Erickson	all classes
E. Norman Flayderman	Civil War pipes
Martin Friedman	meerschaums, Orientals, Africans
William P. "Bill" Fry	regimentals; bibliomania
Lawrence E. Gichner	erotic pipes and tobacco accessories
Burton Gottfried	all classes
Anthony Imbornone	large to supersize meerschaums; selected others
Irving Landerman	all classes; tobacco jars; ephemera
Allan Loeb	meerschaums
Jonathan S. Mopper	regimental bowls
Charles P. Naumoff	all classes
Jay Opperman	meerschaum pipes and cheroot holders
David G. Paegelow	clays
Mario Pancino	regimentals
Herbert G. Ratner	supersize meerschaum and erotic pipes
Michael B. Seiden	meerschaum pipes and cheroot holders
Arthur Silveira	meerschaums
Ralph J. Spiegl	opium and African tribal pipes
John B. Stone	all classes
John B. Sudbury	ethnohistorical & morphological research on clays
Dave Terry	Oriental water, dry smoking and opiums; opium lamps; related ephemera
Eugene B. Umberger Jr.	antique pipes; related ephemera and bibliomania

Chapter 11

RESTORATION AND REPAIR

Returning antique pipes to their former or original condition is fraught with some costly and risky challenges. Amber is expensive nowadays, when it can be found at all in usable lengths. Damaged meerschaums require an artisan with a scrutinous eye and diligent hand to blend an undetectable fill into an already aged, mottled or patinated bowl. These observations are, no doubt, universally known and self-evident. Replacement of parts, as the reader will see, is much easier.

Many pipe shops claim to perform delicate and minute restoration; few ever achieve it and many are abject failures. I have sought out a few reputable craftsmen whose capability and talent qualify them to discuss the subject in sufficient detail so that the ambitious, the compleat collector can restore his pipes at home or in his workshop.

First, I'd like to make my own contribution. J. H. Lowe, Box A-74, Wantagh, New York, 11793, is in its thirtieth year of providing the industry with pipe repair equipment and parts. The 1978–1979 catalog includes nickel, gold and silver bands, wax, fitments, polishing units, taps, stains and rods. Get a copy! You'll find just about anything for the average skilled worker to make repairs.

Got a porcelain bowl which fits too loosely in its reservoir? The housewares department of any retail store which sells canning supplies will stock a blister pack of assorted new corks shaped to fit sundry bottle necks. These can be drilled out and used as replacement sockets. If you need a thinner version or need to enlarge the end of a wood push stem, buy cork gasket, in sheet form, in an automotive supply house; it is available in graded thicknesses and reasonably priced.

Loose or wobbly vulcanite or bakelite stems may be repaired painlessly. There are two accepted methods. In a pan, bring water to a boil. Place the tenon into the water for a few second and then quickly insert the stem into the pipe shank before it cools. As soon as you are satisfied with accurate fit, withdraw the stem, place in cold water, reinsert it in the pipe. If the stem is still loose, repeat the operation for you may not have worked with sufficient speed. The second method is riskier and may ruin the tenon but with care and caution, it can work if the first method has been unsuccessful. Heat the tenon with a wooden or paper match, insert it into the pipe shank and when you are satisfied with a good fit, cool the tenon under cold water.

For a loose or misaligned threaded stem, I recommend teflon thread seal, often called teflon tape, available in any plumbing supply store. Wind the required amount of tape around the screw, use a fingernail to press the tape firmly into the grooves and the tape will adhere securely. A ready substitute, at home, is waxed dental floss or dental tape. More will be said about this aspect later.

Bone and teflon screw replacement is a bit more arduous. One source of supply is Lowe's, the other is Golden Horn Meerschaum Pipes, 1737 S. La Cienega Boulevard, Los Angeles, California, 90035. Screw replacement requires the removal of the old, normally implanted with cement. It can be drilled out or if you choose, apply steam heat to the stem, insert a nail or small diameter drill bit and with properly applied friction, withdraw the old screw while the old cement is pliant. Mount a new screw with a nontoxic cement, making sure of a proper alignment of shank and stem.

Aaron Beck, 1711 No. State Road No. 7, Morell Plaza, Margate, Florida, 33063, carves his own briars and is a qualified repairman whom I heartily recommend for mail order stem repair service. He says: "If the pipe is smokable, I'll repair it." He uses all the proper equipment—lathe, high speed steel cutters, calipers, micrometer, sander, buffer and works principally with delrin, a semi-rigid plastic, to make screws and tenons and marblette, a pheno-resin, known as catalin or bakelite, a synthetic used for costume jewelry. Aaron states that catalin comes in every conceivable color, is stronger, more durable and, most importantly, cheaper than amber for stem replacement. The color Aaron prefers is tortoise shell which resembles a dark amber and blends exceptionally well with patinated meerschaums and antique woods.

Dick Johnson of Johnson's Pipe Shop, 151 So. Hanover Street, Carlisle, Pennsylvania, 17013 is a multi-talented craftsman. He can whittle, hew, carve, sculpt or etch just about any material upon request. Dick likes to work with redolite for replacement stems, a ruby color translucent synthetic. He has replaced several stems for me, fitted them snugly into the accompanying etui and charged a nominal sum. I've met Dick and he is, by reputation and performance, all I've described. He has also made amber stems for clients who insist on ne plus ultra. Dick responds well to mail order repairs. Parenthetically, he discreetly markets antique pipes to interested collectors and always has a large selection of choice items for sale.

Hornepipe's Bob Horne moved from California in summer 1979, but he promises to continue his handicraft after he is relocated to Route 1, Box 82H, St. Maries, Idaho, 83861. Bob Horne is a silver- and goldsmith who comes highly recommended by Doug Murphy, Tinder Box International Ltd. Bob's specialty is repairing cracked shanks on antique pipes. Bob can mount a 14K or sterling silver band to just about any round, oval, diamond or square shank. In refitting, Bob is prepared to furnish hand-made screws and stems. For shanks which are completely cracked, Bob's skill is applied to a free-form band to cover the entire mended break. And in the very worst case, should the shank be completely broken off, he repairs the pipe to its former functional state and crafts a mid-shank band to conceal and reinforce the break. For those antique meerschaums and briars with original factory bands, Bob promises to preserve and extend it. Hornepipe offers a variety of gold finishes—plain, hammered relief and full nugget, proposes a design and provides a cost estimate for approval.

So much for repairs. The care, cleaning and conservation of antique pipes can be accomplished with little investment of money, much investment of time. The following commentary has been furnished by Doug Murphy who deserves a hearty round or two of applause for divulging his trade secrets. He is a white collar executive who is enthralled with this blue collar chore. I have edited Doug's monologue but the ideas are his. The Tinder Box and C.A.O. products are Doug's recommendations. Since they serve the purpose intended, so be it! There are certainly a host of products in the marketplace which may substitute for those Doug proffers and I leave it to the reader to make that decision. I disagree slightly with one of Doug's procedures and it pertains to polishing metal. I believe that a patina on metal enriches and enhances an antique pipe and I do not polish lids, collars, bands or ornamentation—frequent polishing may remove any plating, scant engraving or chasing. The balance of this chapter belongs to my compatriot, Doug Murphy.

SUPPLIES AND TOOLS NEEDED

Cotton balls
Cotton swabs (Q-Tips, paper stem only)
Clean cotton knit cloth (old "T" shirt)
Plastic drip cloth or newspaper
Tinder Box Pipe Sweetener*
Alcohol
C.A.O. Meerschaum Cleaning Fluid**
C.A.O. Meerschaum Magic Care Cloth**
C.A.O. Meerschaum Antiquing Compound**
Brasso metal polish
Soft electric toothbrush (brush part only) or soft child's toothbrush
Soft, medium and hard adult toothbrushes
Bristle pipe cleaners (Ream 'n Klean)**
Thick tapered pipe cleaners**
Dunhill or Savinelli Mouthpiece polish**
Dunhill or Savinelli Pipe Bowl polish**

Most antique pipes that I have encountered could use cleaning. Some need only light dusting and polishing while others are very soiled and require thorough cleaning. Antique pipes, like antique furniture, develop a character of their own through years of use. Whenever I clean an antique pipe, I try not to lose its character and charm. If a Chippendale chair has been cleaned and restored correctly, it will still possess all the character of its years. It would be a shame if someone took a Chippendale, sanded, stained and polished it until it looked like a copy. And so it is with a pipe; even if I have a badly damaged pipe, I only remove that part of its character which would otherwise detract from it.

The first step in cleaning and restoring an antique pipe is an appraisal of the work to be done. The following are questions you should ask yourself before you start work:

1. What type of stem does the pipe or cigar holder have—is it amber, cherrywood and horn, horn, bakelite, redolite or amberoid?
2. Does the stem need cleaning or repair?
3. Does the pipe or cigar holder have a metal cap or shank fitting? If so,

*Available in The Tinder Box Pipe stores
**Available in The Tinder Box Pipe stores and other tobacconists

what type of metal is it—gold, silver, brass or chromium?
4. How much cleaning does the bowl require—light, medium or heavy?

Cleaning and Repairing Stems

Since I always start with the stem, I'll address cleaning and repairing horn and cherrywood and horn stems first. I clean the inside of the stem with bristle pipe cleaners (Ream 'n Klean) moistened in Tinder Box pipe sweetener. This is not a plug for Tinder Box, but most other sweeteners on the market do not have enough solvent power to cut through years of tar build-up in the stem. The alcohol content of this sweetener also helps to sanitize the stem if the pipe is to be used again. Next I replace any worn parts such as corks. Sometimes, it is necessary to shim the threads at the joint of the mouthpiece and stem. Shimming these threads is as simple as wrapping the appropriate amount of teflon tape around them. This type of repair is simple to do and certainly adequate for a pipe that is only to be displayed. If you intend to use the pipe regularly, I recommend a new stem as the permanent resolve, teflon tape as only a temporary fix. The mouthpiece or other parts of the stem may be replaced on a pipe to be displayed, as sometimes, they are missing and at other times they are bug eaten from improper storage. Whenever possible I like to use the original stem and mouthpiece.

If it is an amber stem, I caution you to be very careful. With age, amber becomes very brittle and cannot withstand much stress—if an amber stem is forced a little too much in trying to free it from its bone screw fitting, it could break. It can break when dropped and harsh cleaners may stain it. I clean the inside of an amber stem with a bristle pipe cleaner (Ream 'n Klean) lightly moistened in pipe sweetener. You must take care not to get sweetener on the exterior. I first clean and shine the exterior with C.A.O. Meerschaum Cleaning Cloth and then polish it with cotton balls. This will give it a nice luster. If the stem is scratched or has teeth marks, do not try to grind, file or buff these off by hand or on a buffing wheel. This will certainly cut through the beautiful outside finish. A little too much heat or one careless move could cause a beautiful old amber stem to shatter.

Bakelite, redolite and amberoid can all be cleaned on a buffing wheel with stainless steel buffing compound, if you have a buffer. If you do not, both Dunhill and Savinelli manufacture a stem polish which can be applied by hand and will do the job nicely. The inside of these stems can be cleaned with bristle cleaners (Ream 'n Klean) moistened in pipe sweetener.

Cleaning Pipe Caps and Shank Fittings

I only use one metal polish for all types of metal fittings—Brasso. At the mention of polishing metal fittings on an antique pipe, I am sure many of you envision ruining the character of the pipe. But, if the pipe had been in use all these years, I am sure the metal would not be tarnished and soiled. Let me return to my original example of the Chippendale chair. If you owned a Chippendale chair that was dirty, wouldn't you use a furniture cleaner? Or, if the finish were chipped or blistered, wouldn't you remove the old and damaged finish and apply a new one? Just as a professional furniture refinisher would carefully remove and renovate the finish on antique furniture, I clean and restore the metal fittings on pipes!

I am careful not to remove all the pipe's character by cleaning every nook, cranny, dent and scratch, but I do clean and polish.

I use a cotton swab (Q-Tip) moistened in Brasso. This is a tool soft enough not to scratch but firm enough to get the job done. Swabs reach hard to get places. After the metal has been cleaned, I wipe away dried and excess polish with a cotton cloth, take a cotton ball and give the metal a final buff. This works equally well on silver, gold, brass or chromium.

Cleaning Meerschaum Pipes

I want to speak about cleaning meerschaum pipe bowls first because they seem to hold the most mystery. You must understand something about how meerschaum pipes color to understand how to clean them. Meerschaum pipes are coated with wax and it is the wax that colors. So, if you remove the wax, you remove the color. Therefore, any cleaning you do must leave the wax intact. I am confident there are some who have handled, polished and possibly even cleaned a meerschaum bowl with harsh cleaner which has had no effect on the color of the pipe. This occurs because, more than likely, what they have is a pre-colored meerschaum, one that has been artificially stained.

Before I start cleaning the pipe, I first assess the work to be done. Is the pipe colored or almost white? How dirty is it? Is the pipe naturally colored or pre-colored? After considering these questions, I decide which of the following cleaning methods I will use:

1. If the bowl is naturally or artificially colored and only needs light cleaning, I brush on a coat of C.A.O. Meerschaum Cleaning Fluid with the applicator brush provided. After a few minutes, I wipe the bowl with a C.A.O. Magic Care Cloth. This process will give the bowl a nice shine. To improve upon this shine further, I polish with cotton balls and, for hard to get at areas, as on full relief carved pipes, I use a cotton swab.

2. If a naturally or artificially colored bowl needs what I consider medium cleaning, I start by brushing on a coat of C.A.O. Meerschaum Cleaning Fluid. While the cleaning fluid is still wet, I take a very soft toothbrush, children's soft or electric toothbrush type, and lightly brush the indented, fluted or carved areas to remove the dirt and grime collected over the years. To clean this same dirt from the smooth areas of the bowl, I brush on a coat of C.A.O. Meerschaum Cleaning Fluid and while it is still wet, I take a cotton ball saturated with C.A.O. Meerschaum Cleaning Fluid and rub. The dirt will lift from the surface of the bowl much like cleaning a leather shoe with saddle soap. I change cotton balls as often as necessary and I do not continue using one that has become soiled. Next, I allow the bowl to dry and brush on one more coat of Meerschaum Cleaning Fluid. I let the bowl dry again, wipe with a C.A.O. Magic Care Cloth and finish by polishing with cotton balls and cotton swabs.

3. Naturally colored or artificially colored meerschaum bowls needing heavy cleaning require several different approaches. Sometimes a pipe

or cigar holder that appears very hard to clean may prove easier than you think. What appears to be hard dirt is often only a thick coat of dirt mixed with grime which has collected on the pipe in years of storage. These bowls can be cleaned using the medium cleaning process; I repeat any of the steps as often as necessary. I always start by trying these steps first and often I am pleasantly surprised. If the dirt on the pipe does not lift out, then I go one step further. In the past, I have used alcohol as a surface cleaner. I am very careful in its application. An artificially colored meerschaum can almost always be cleaned with alcohol and it will not cause damage to the color, but I always proceed cautiously. I do not saturate a cloth with alcohol and start scrubbing. I start by moistening a cotton swab with alcohol and test the surface in some inconspicuous spot. Many times this alcohol moistened cotton swab will be all I need to clean the bowl. If I need to clean larger areas I use a cotton ball moistened with alcohol. I always test first with a cotton swab and then progress to a cotton ball. And, I never use so much alcohol that it runs onto the pipe bowl when I press either the swab or the ball against the pipe wall. Again, I use alcohol sparingly.

You can use alcohol to clean an unsmoked or lightly smoked meerschaum bowl. Sometimes, however, cleaning a white meerschaum pipe or cigar holder soiled by handling or improper storage can be the most difficult cleaning job. Any alcohol you apply will remove some wax. Obviously, the more alcohol you use, the more wax you remove. Removing some wax is not that important because the pipe has no color yet. As in cleaning an artificially colored meerschaum pipe, always use caution with alcohol. Start using a cotton swab moistened with alcohol and test; then progress to the cotton ball if necessary. Alcohol will clean almost any surface dirt and will remove some stains. If the meerschaum itself has been stained by ink, oil or mold, as examples, I am afraid there is no solution. There are other procedures I know and have tried but none removes all the stain, some not sufficiently to bother with the application.

Cleaning a naturally colored meerschaum is not very difficult if it needs only light or medium cleaning. You can use either the light or medium cleaning I described without any worry of damaging the beauty. If you use a toothbrush, be sure it is the softest you can find and use it lightly. If the pipe needs heavy cleaning, your job may be more difficult. As with meerschaums I described earlier, I always start cleaning a bowl that appears to need heavy cleaning by using medium cleaning. I will repeat any and all these steps as often as necessary; many times, I get the pipe clean using only these steps. Sometimes the dirt is stubborn and cannot be removed with these techniques.

Before I explain the next cleaning processes, you must understand a little more about the wax used on meerschaums. There are two types of wax. The first is old wax, a combination of beeswax, sperm whale oil and tallow. This wax is harder than you think. New wax consists only of several coats of beeswax. Both waxes can be cleaned with alcohol, but read the following thoroughly before trying to use alcohol on a colored meerschaum bowl.

Old wax is pretty hard and most of the time can be cleaned with alcohol; but, as with other meerschaums, there is always the possibility that alcohol will damage the finish. There are two ways to minimize failure. One, decide whether the pipe in its present condition is so damaged and unattractive that it is worth a try and maybe a failure. The second way to minimize failure is to test some inconspicuous spot on the pipe before you proceed to use alcohol on the entire bowl. I have had good results using an alcohol moistened cotton swab to clean dirt which rejected other cleaning techniques. *Never* soak the cotton swab with alcohol because as soon as you press it against the bowl, alcohol will run and collect on the bowl's finish. Doing this is asking for trouble. After you have softened and cleaned most of the surface dirt with alcohol, revert to the basic steps for a bowl needing medium cleaning. Follow the steps for medium cleaning and you should have a pipe or cigar holder that is clean and attractive once again.

A pipe with new wax has a much softer surface than one with old wax. New wax is the only type found on today's meerschaum pipes, but you will also find that some antique meerschaum pipe and cigar holders have this softer finish. Any amount of alcohol you apply to this finish will remove some wax. If you are not careful and allow the alcohol to run or collect on the bowl, there will appear a whitish stain that will not come off. You can still clean the bowl with alcohol if you are careful. Imagine a car whose paint is badly oxidized. There are two solutions to make the car shine again: the first, use rubbing compound to remove the oxidized paint and if you rub too hard for too long, you will rub through to the primer. So what? The second alternative was to repaint the car!

In summary then, for meerschaum bowls requiring heavy cleaning, first use the steps described earlier for a bowl needing medium cleaning and repeat the steps if necessary. Only use alcohol as a last resort on a pipe so soiled that it is unattractive in its present state anyway.

Porcelain Pipes

Stems on antique porcelain pipes are, for the most part, like those found on antique meerschaum pipes. You will find porcelain bowls that fit into either porcelain bottoms or horn bottoms. Stems will be horn or cherrywood and may have a flexible hose attached to the mouthpiece. Clean these stems as you would restore the stem of a meerschaum pipe.

Porcelain pipe bowls are often fitted with caps and shank fittings of silver, brass or chromium. As on meerschaum bowls, I clean these fittings with Brasso but only after I clean the bowl. Most porcelain bowls only need light or medium cleaning since porcelain is hard, does not stain easily and is easy to clean. There are two methods I use to clean them. The first is to use a household glass cleaner on the surface and wipe dry with a soft cloth. This works quite well, especially if the pipe is only to be displayed. If the pipe is very soiled or I wish to smoke it, I wash it and the porcelain moisture trap, if there is one, in warm water with a mild dishwashing liquid. Bottle brushes and bristle pipe cleaners (Ream 'n Klean) can be used to clean the inside of the bowl and other hard to get at places. After the pipe has been rinsed and dried, I clean any metal fittings.

The pipe is now ready for reassembly. Many times the cork fittings that hold the bowl and stem in place are worn and useless. I replace the cork, shaping it into a cone, overlapping it slightly and gluing it together. The next step in reassembly appears to be decorative but it is also functional. I refer to the cord and tassels that accompany many long stemmed pipes. These serve to keep the bowl, reservoir and stem from disassembling, falling and breaking. I like to use the original cord and tassels whenever possible. If they are missing or damaged, I can find cord and tassels at most yardage stores; with very little work, these can be fashioned into nice replacements.

Wood Pipes

Wood pipes come in a wider variety than any other type. There is cherry, oak, briar, walnut burl, elm burl and rosewood, just to name a few. They are found in various styles from the long Austrian to the contemporary briar and their stems represent a large variety of materials. Long horn and cherrywood stems like those on meerschaum and porcelain can be cleaned and restored as I described earlier. There is also amber, redolite, bakelite and vulcanite. Amber is very fragile and you should clean the inside carefully with pipe sweetener and the exterior with a C.A.O. Magic Care cloth. Polish redolite, bakelite and vulcanite stems on a buffing wheel with stainless steel compound. If you do not have a buffer, clean such stems by hand with either Dunhill or Savinelli mouthpiece polish.

Generally, wood is a very durable material and is not difficult to clean. As with other pipes, the first step in cleaning a wood pipe bowl is to decide how much work is needed and to what condition do you desire to restore the pipe. Because it can be done, I will elaborate on how you can completely refinish a wood pipe bowl. Personally, as with the Chippendale chair, I like to stop short of refinishing, whenever possible, and only concern myself with restoring the pipe to an acceptable condition.

If you intend to smoke an antique wood pipe, start cleaning from the inside out. Ream out the carbon in the bowl. Do not use a knife or other makeshift tool—use a pipe reamer! Next, clean the inside of the shank and stem with sweetener and bristle pipe cleaners. Moisten the cleaners in sweetener and run them in and out of the shank. Unless the draft hole is very small, I have found that two at a time work best. Clean the inside of the stem with bristle pipe cleaners and sweetener. Following this, clean any metal caps or shank fittings as needed with Brasso.

Now to the outside of the bowl. If the pipe only needs light to medium cleaning, the first step is to clean the outside with alcohol. On smooth bowled pipes, use either a cotton ball or soft cloth moistened with alcohol. On rough finished pipes or on carved areas of predominantly smooth pipes, use a cotton swab or soft toothbrush to clean the hard to get at places or stubborn dirt. The difference in the application of this technique for a pipe needing light and a pipe needing medium cleaning is only the amount of time required to get the bowl clean. After you clean, the pipe will have a dull appearance. Give a smooth pipe a nice luster by buffing it on a polishing wheel with carnuba wax or by hand using Dunhill or Savinelli pipebowl polish. Rough or carved bowled pipes cannot be polished on a wheel and must be polished by hand. To repolish a rough pipe, give it a good scrub with a medium bristle toothbrush; if this is not an adequate shine, apply some Dunhill or Savinelli bowl polish. After applying polish, brush with a soft bristle toothbrush, then buff with a soft cloth. These techniques will clean and restore 99% of the woods.

I have encountered a wood pipe or two so soiled and damaged that it required heavy cleaning. When I speak of heavy cleaning, I mean refinishing. I do not like to refinish a pipe totally, but unfortunately, sometimes a pipe is in so sad a state of disrepair that refinishing is the only solution. I start by cleaning out the grime with a cotton ball or cloth saturated with alcohol. Next, I use a soft cloth saturated with acetone. Use rubber gloves if you have sensitive skin, do not breathe the fumes and do not work near a flame. Acetone should remove the old finish as well as clean any remaining dirt. Better quality wood pipes have a natural finish but some have been varnished or shellacked. If your pipe has a varnish or shellac finish, you may have to make several applications of acetone or use a stiff toothbrush saturated with acetone. The acetone, then, will remove dirt, finish and some color or stain. If you want to recolor the pipe, touch up the scratched or damaged areas and stain the pipe. Pipe stain is not the same as that for furniture. Most furniture stain is oil based to protect the wood against damage by sealing it. You do not want to seal a wood pipe bowl since its porosity aids smoking. Therefore, stain a wood pipe with an alcohol base stain. Commercially available alcohol base wood stains and leather dyes may be used. After you have cleaned, finished and restained, the next step is to repolish. This can be done by following the light to medium cleaning procedure.

I have explained only some of the basics. I have not discussed cleaning and restoring all types of pipes nor all the cleaning techniques for those about which I have spoken. I am quite sure that by the time this is printed, I will have learned additional techniques to clean and restore and will have gained more of a very important aspect in doing anything—experience. Learning is a continuum of the past, present and future. So, before you try a cleaning technique that is new to you, experiment first on a pipe you hold less dear. Use common sense when handling chemicals, read the directions. And, since your objective will be to restore and refinish already valuable and irreplaceable antique pipes to a new condition and improved appearance, a word to the wise—be persistent. . .be patient. . .and prudent.

As a parting word from the author, I commend to the more skilled or to the more daring collector two textbooks which elaborate on aspects of repair and restoration not covered in this chapter. Sadly, both books are now out-of-print but may be found in an antiquarian bookshop. The first is Carl Avery Werner's TOBACCOLAND, Tobacco Leaf Publishing Company, New York, 1922, which details, among other things, how to make stem cement, repair amber and pipe cases and one method of artificially coloring a meerschaum. Emil Doll's THE ART AND CRAFT OF SMOKING PIPES, H. Behlen & Bros., New York, 1947, is dedicated to pipe manufacturers, jobbers, wholesalers, retailers, salesmen and smokers everywhere. It is, in the truest sense, a bible for the wood pipesman for it contains such exotic information as a formula for mastic putty, the techniques of stain, varnish and shellac, bit and ebauchon standards and sizes and a recommended list of essential materials for wood finishing.

Chapter 12

RECOMMENDED READING

With Pipe and Book

"With Pipe and Book at close of day,
O! what is sweeter, mortal, say?
It matters not what book on knee,
Old Izaak or the Odyssey,
It matters not meerschaum or clay.

And though our eyes will dream astray,
and lips forget to sue or sway,
It is 'enough to merely Be,'
   With Pipe and Book.

What though our modern skies be grey,
As bards aver, I will not pray
For 'soothing death' to 'succour' me,
But ask thus much, O! Fate, of thee,
A little longer yet to stay
   With Pipe and Book."

   Richard Le Gallienne, 1889

Of the thousand odd tobacco volumes in my own library, in those of the George Arents Jr. Tobacco Collection of the New York Public Library, in the Tobacco Merchants Association of the United States' Howard S. Cullman Library, New York City, and the Robert Hays Gries Tobacco Collection of the John G. White Department of Folklore, Orientalia and Chess, Cleveland Public Library, Ohio, few books in English stand out as essential, illustrated and in-depth surveys of this collecting field.

One principal disadvantage for the American pipe collector is that European literature covers the field admirably well, all things considered. Therefore, I beseech the serious collector to obtain those books which best address his peculiar pipe collecting subset including any foreign language illustrated books described herein. Many of these foreign titles are readily available in retail outlets which traffic in Old World books or can be ordered directly from the authors.

For the generalist, the two earliest accounts in the English language and, without question, the most authoritative and priceless compendia of tobacciana ever assembled were the Bain Catalogue and the Bragge Notebook. Through time, the former has been obscured and overshadowed by the arch collector's manuscript which detailed each of the more than 6,000 specimens he collected. Owning the original or a facsimile of both should be the object of every serious collector interested in the smoking implements of all nations! BIBLIOTHECA NICOTIANA; A FIRST CATALOGUE OF BOOKS ABOUT TOBACCO, privately printed by Josiah Allen, Birmingham, England, in 1874, is an annotated bibliography of the

169 books on tobacco William Bragge had collected to that date. I quote verbatim two entries from this bibliography:

> TOBACCO: ITS HISTORY AND ASSOCIATIONS, USE AND ABUSE, INCLUDING AN ACCOUNT OF THE PLANT, AND ITS MODES OF USE IN ALL AGES AND COUNTRIES, SHEWING IT TO BE THE SOLACE OF THE KING AND THE BEGGAR. Comprising Prints and Woodcuts; Portraits of renowned Smokers; Tobacco Papers; Numberless Cuttings and Extracts; Pipes, Cigars, Snuff, and Snuff Boxes, and all the Smoker's Paraphernalia; Statistics of Consumption, Revenue, etc. in Relation to this Wonderful weed, and in fact, every conceivable item of interest that could be gathered in relation to the subject. The result of over thirty years labour in Collecting, mounted and arranged in 10 large folio Volumes (with specially printed title pages), elegantly bound in half green morocco extra, gilt tops, by A. W. Bain. 1836.

> ILLUSTRATED CATALOGUE OF A COLLECTION OF PIPES OF ALL AGES AND COUNTRIES: Pipe Cases, Tobacco Stoppers, Tobacco Pouches, Cases, and Jars; Snuff Mills, Snuff Graters, Snuff Bottles, Snuff Boxes, Snuff Jars; Flint and Tinder and other Fire Strikers, and Lighting Apparatus and objects connected with the use, in any form, of Tobacco. MS.

The first entry is the Bain monograph. The second manuscript is Bragge's and it evolved into the Notebook which he published at his own expense in 1880 as: BIBLIOTHECA NICOTIANA: A CATALOGUE OF BOOKS ABOUT TOBACCO, TOGETHER WITH A CATALOGUE OF OBJECTS CONNECTED WITH THE USE OF TOBACCO IN ALL ITS FORMS, Birmingham, England. The J. Trevor Barton Collection, sold to the US Tobacco Company, included a copy of the Bragge Notebook.

In their absence, a very good substitute is HISTORICAL AND ETHNOGRAPHICAL SMOKIANA by Robert T. Pritchett, 1890. This tome includes 101 illustrations, 41 of which are hand-sketched authentic portraits of the pipe family from the Sixteenth Century to that day. THE PIPE BOOK by Alfred Dunhill, 1924, should appeal to not only the pipe collector but also the pipe smoker; it contains 28 full page plates, 230 illustrations and relates the legends and customs associated with pipe smoking. His son, Alfred H. Dunhill, revised his father's unrivaled book in 1969, inserting minor textual changes and some new material. In July 1971, this scholar and pipe enthusiast died at the age of 76. Since that time, only one other English language book has been published which acknowledges the existence of the collectable antique pipe. In 1974, Carl R. Ehwa Jr., now a master blender for the McClelland Tobacco Company, Kansas City, Missouri, wrote the spectacularly illustrated BOOK OF PIPES AND TOBACCO. The photography is superior and the presentation is typical of the fine books printed in Italy. In "Part Two, The Amiable Pipe," Ehwa does justice to a variety of antique pipes from the Ehrlich and Half-and-Half Collections.

In foreign languages, there are ample general works to choose from and the authors are accredited authorities on the subject. Among those in French available today, the foremost work is LA PIPE, written in 1973 by André Paul Bastien,

Director of a tobacco journal, *Revue des Tabacs* and its adjunct publication, *Pipe Magazine*. A. P. Bastien is a prolific writer of fiction as well, but in the nonfiction LA PIPE, he has assembled lush color photographs of choice pipes, used simple French grammatical construction and excelled in layout and documentation. A parallel volume appeared two years earlier but covers a broader range of tobacco artifacts as the title implies: LES OBJETS DU FUMEUR by Michel Belloncle. This book is part of a series entitled Collection de l'Amateur. The photographic coverage is predominantly black and white, but the illustrations are excellent and will broaden the perspectives of a collector who may desire to collect ancillary utensils as snuff boxes and rasps, match strikers and tobacco boxes.

The newest entry in French belongs to Jean-Charles Rhein, a Swiss tobacconist. At his shop Tabac Rhein, the Rhein family has been selling tobacco products since 1905, but it has been the pipe collection which now draws daily crowds—the collection has been on display since 1961. L'ART DE LA PIPE is an exquisite and very expensive catalog (50 S.fr), printed in 1978 in conjunction with a Rhein exposition organized by Cabinet d'Amateur in Geneva. This is no ordinary catalog, for the professional color photography is accompanied by brief biosketches which reveal pipe dimensions and provenance—that makes it invaluable to any researcher and collector.

Among the foremost Italian pipe collectors is Giuseppe Ramazzotti, a scholar, distinguished biologist, international authority on tardigrades, pipe collector and once-managing-director of the defunct *Il Club della Pipa*, whose reviews by the same name are currently in great demand. In 1967, he authored one monograph on the history and development of the pipe: INTRODUZIONE ALLA PIPA, and in 1946, co-authored another with Dino Buzzati: IL LIBRO DELLE PIPE, reprinted in 1966. Both are distinguished and literate accounts but lack illustrative depth. Another Italian book, however, from the distinguished family of Rizzoli art books, does warrant mention—Diego Sant'Ambrogio's PIPE, 1966. This small handsome book in a slipcase is a full color panorama, page upon page of both antique and modern pipes in exotic settings; the text is minimal and interwoven with black and white hand-drawn sketches of pipes.

For those who read German, I submit the following for consideration. A friend, Helmut Hochrain, is a freelance writer whose literary forté is a thorough knowledge of beer, wine and tobacco. He neither smokes a pipe nor collects them; he is, however, enchanted with their mystique. To prove that, he has entertained the German reading public with three books on tobacco and smoking utensils. DAS GROSSE BUCH DES PFEIFENRAUCHERS, published in 1973, is the most richly illustrated tome. Herr Hochrain's use of sepia tone illustrations and photographs, perhaps intended as tobacco brown, is most effective; so for those who are wont to plod through German text, there is satisfaction in knowing Herr Hochrain writes with authority and erudition. The collector can also avail himself of A. P. Bastien's LA PIPE, translated in 1972 into German as VON DER SCHOENHEIT DER PFEIFE or Sant'Ambrogio's PIPE, translated in 1967 as PFEIFEN.

Georg Brongers, curator of Niemeyer's Nederlands Tabacologisch Museum, is the local Dutch expert. In 1964, Mr. Brongers had written NICOTIANA TABACUM, an account in English which covers the realm of tobacco culture, smoking, art and collectables; there are quality photographs of occasional pipes. The criticism American collectors would render is that the book is written about tobacco developments in the Netherlands—it is still in print. For those who read Dutch, G. Brongers gifted the reader-collector world with VAN GOUWENAAR TOT BRUYÈRE PIJP in 1978. With only slight disappointment, I reviewed the book to find that although *gouwenaar* is Dutch for churchwarden and *pijp* is the Dutch word for pipe, the book was not as encompassing or inclusive of pipes and their ethnography as I had hoped for.

In Scandanavia, one tobacco merchant stands out among his peers— Ole W. Ø. Larsen. Ole is a congenial retailer, a knowledgeable collector and his briars are noted for their quality. In conversation with me during 1978, Ole announced the publication of GAMLE PIBER as one of a series of six collector books on interesting Danish antiques. In it, Ole describes antique pipes still within the reach of Scandanavian middle-class incomes and shows a variety of antique meerschaums from his museum collection. My hat is off to Ole and GAMLE PIBER!

In pipe subset specialties, there are many authoritative works and I shall mention the most outstanding. To begin, I set aside any pursuit by the eager meerschaumers by stating emphatically that there is no text in print today dedicated wholly and solely to that pipe. As to briar, just about any tobacco book will mention this material as an excellent smoking pipe and as one of tomorrow's collectables worthy of investment, but no particular book, in my view, is worthy of special merit.

If you collect Oriental water, dry-smoking or opium pipes and read Japanese, visit the U.S. Library of Congress' Orientalia Division and review NIHON NO KITSUEN-GU a non-commercial outsize book, published in 1966 under the supervision of Nihon Sembai Kosha, the Japanese Tobacco Monopoly Corporation, Tokyo. The title of this quasi-governmental non-trade text is translated as PIPES AND TOBACCO ACCESSORIES. This is a superior volume which contains enough photographs to capture the eye and the imagination of any prospective collector and induce, perhaps seduce him to specialize. More easily, one can whet his appetite by viewing the many Orientals photographed in the Parke-Bernet PB 84 Sale 615 auction catalog or obtain a copy of a small but exceptionally intriguing gratuitous brochure entitled *Smokin' Clean* recently released by the Japan Gallery, 1210 Lexington Avenue, New York.

Into ethnographica of the American Indian? The annual report of the Board of Regents of the Smithsonian Institution for the year ending June 30, 1897, Report of the U.S. National Museum, Part I, published by the Government Printing Office in 1899, contained Joseph D. McGuire's 296 page monograph, "Pipes and Smoking Customs of the American Aborigines"—with, I add, 239 illustrations. George A. West, a successful Milwaukee attorney and an ardent collector of American Indian pipes, shared his knowledge of pipes by frequent contributions to *The Wisconsin Archaeologist*, the journal of the Wisconsin Archaeological Society, starting in 1905. He donated his collection to the Milwaukee Public Museum in 1913 and the Museum published a compendium of his monographs in 1934. Since that time, this two volume thesis has been recognized as the leading reference and was in so much demand that it was reprinted in 1970 by Greenwood Press as TOBACCO, PIPES AND SMOKING CUSTOMS OF THE AMERICAN INDIANS. West's 995 page work is replete with descriptive terms and contains 257 black and white plates. "Memoir Number 5" of the Missouri Archaeological

Society, published in September 1967, is a tribute by the President of that Society to the Indians of the State. Titled TOBACCO PIPES OF THE MISSOURI INDIANS, this is a brief but detailed account of the pipes peculiar to that region, accompanied by an excellent bibliography. In 1977, J. C. H. King, Assistant Keeper, Department of Ethnography, Museum of Mankind, British Museum, prepared a softcover booklet to describe 100 select pipes from the British Museum Collection. This slight monograph is SMOKING PIPES OF THE NORTH AMERICAN INDIAN and contains 28 black and white plates.

Two additional works are fascinating anthropological and historical accounts of New York State Indian pipes, requiring, I confide, a college degree just to understand the titles. The earlier is NOTES ON THE AMERIND MANUFACTURE OF SMOKING DEVICES AS ARTISTIC EXPRESSION IN NORTHEASTERN IROQUOIA, written in 1968 by Anthony L. Sassi, Jr.; the other is SMOKING TECHNOLOGY OF THE ABORIGINES OF THE IROQUOIS AREA OF NEW YORK STATE by Edward S. Rutsch published in 1973. The authors focus on the Iroquois, their artifacts, the use of tobacco, and both books are amply illustrated. As I finalized this manuscript, I discovered that the Smithsonian Institution had just published a pictorial survey by John C. Ewers, a noted anthropologist, entitled INDIAN ART IN PIPESTONE.

As to African ethnographica, one little known opus is PIPES DU CAMEROUN which details a collection of clay effigy pipes in the collection of Mont Febe Benedictine Monastery, Yaoundé, Cameroon. The authors, P. Luitfrid Marfurt, a monk, and Jean Susini, the photographer, executed this fully documented treatise in 1967. It includes black and white close-up photographs to portray the myriad expressions of African craftsmanship and imaginative sculpture. In this volume, pictures outnumber words!

The collector of porcelain pipes and that even smaller population who specialize in reservistenpfeifen [regimentals] must patiently await the first effort by an American collector and friend, William P. "Bill" Fry of Riverside, California. Bill need only muster sufficient fortitude and financial support and he will share his knowledge of pre-World War I Imperial German military units, traditions and customs. Therein lies the story of reservistenpfeifen!

I leave to last the most documented and illustrated field, the clay. The archaeologist, not the collector, deserves the credit. In the near distant past, a large number of booklets and brochures has come off the presses with great frequency, in limited numbers and without much notariety. The majority are English works and their titles reveal the contents: Adrian Oswald, ENGLISH CLAY TOBACCO PIPES, 1967 and CLAY PIPES FOR THE ARCHAEOLOGIST, 1975; David R. Atkinson and Adrian Oswald, LONDON CLAY TOBACCO PIPES, 1969; David R. Atkinson, TOBACCO PIPES OF BROSELEY SHROPSHIRE, 1975 and SUSSEX CLAY TOBACCO PIPES AND THE PIPEMAKERS, 1977; Laurence S. Harley, THE CLAY TOBACCO-PIPE IN BRITAIN, 1963, reprinted with addition, 1976; R. J. Flood, CLAY TOBACCO PIPES IN CAMBRIDGESHIRE, 1976; Iain C. Walker, THE BRISTOL CLAY TOBACCO-PIPE INDUSTRY, 1971 and CLAY TOBACCO-PIPES, WITH PARTICULAR REFERENCE TO THE BRISTOL INDUSTRY, 1977, a four volume set; R. G. Jackson and R. H. Price,

BRISTOL CLAY PIPES: A STUDY OF MAKERS AND THEIR MARKS, 1974; Jenny E. Mann, CLAY TOBACCO PIPES FROM EXCAVATIONS IN LINCOLN, 1970–74, 1977; David Helme, THE CLAY TOBACCO PIPE: AN ILLUSTRATED GUIDE, 1978; and the latest contribution edited by Peter Davey, THE ARCHAEOLOGY OF THE CLAY TOBACCO PIPE, I. BRITAIN: THE MIDLANDS AND EASTERN ENGLAND, 1979. These works represent extensive and intensive research in a field which heretofore clay collectors claimed there was none. Their prayers have indeed been answered.

I also acknowledge, with the greatest pleasure, two friends in Europe who have joined the publishing world of clay literature. While Don Düco is pursuing a degree in museology, he has prepared *Jaarverslags*, annual reports for 1975–77, compilations of articles, books and assorted literature on tobacco pipes in which he sketches, with a fine hand, his clay pipe accessions. The annuals are for sale at Pijpenkamer Icon. Jean-Léo, once a reporter in World War II, now resides in Brussels where he has international acclaim as a noted bookseller, clay pipe collector and bibliographer. His contribution on clays is a 1971 limited edition illustrated work entitled LES PIPES EN TERRE FRANÇAISES DU 17ᵐᵉ SIÈCLE à NOS JOURS. With diligence and infinite detail, Jean-Léo documented the French clay pipe industry and included facsimile catalog pages from famous clay manufacturers, Gambier, Job Clerc, Duméril, inter-alia . . . an essential, visual reference for the Western European clay collector. The last entry is Jean Fraikin's 1978 brochure, LA FABRICATION DE LA PIPE EN TERRE. As curator of le Musée de la Vie Wallonne, Liège, M. Fraikin has given the reader insight into clay pipe manufacturing processes of disestablished Belgian firms, adding depth with some very fine representative illustrations of plant layout, workers' benches and catalog facsimiles.

As a final word, I have reviewed my own library and notice that my European holdings are more colorful than their English and American counterparts. Without supportive literature, my pipe collection would have been a mere accumulation of just so many pipes of varied origins. With these books, I have acquired understanding and a greater appreciation of pipes. No collector can become thoroughly familiar with his holdings without research; I am fortunate for I am not limited to my native language for that research. To every collector, I submit that he who decries the lack of information need only acquire any of the pertinent foreign books listed in this chapter and gain the desired knowledge. Most other works on tobacco may devote a paragraph, page or chapter to the pipe but do not give it the descriptive attention needed by collectors eager to understand the historical perspective. A picture is still worth a thousand words, perhaps more in a foreign language you have not mastered. Parenthetically, the majestic and mind-boggling pipes displayed in them may never be for sale in our lifetime. In absence of the real thing, why not own its image? And, I am prepared to furnish bibliophilic assistance to anyone who seeks to build a library of relevant literature, foreign or domestic!

SELECTED BIBLIOGRAPHY

Books

An Old Smoker. *Tobacco Talk*. Philadelphia: The Nicot Publishing Company, 1894.

Aoe, Shū. *Satsu-Gu Tabako Roku*. Tokyo: Zenschichi Maruya, 1881.

Apperson, G. L. *The Social History of Smoking*. New York: G. P. Putnam's Sons, 1916.

Aschenbrenner, Helmuth. *Tabak von A bis Z*. Bremen: Martin Brinkmann AG, 1966.

Atkinson, D. R. *Sussex Clay Tobacco Pipes and the Pipemakers*. Eastbourne: Crain Services, 1977.

―――. *Tobacco Pipes of Broseley Shropshire*. Essex: Hart-Talbot, 1975.

Atkinson, David and Oswald, Adrian. *London Clay Tobacco Pipes*. Oxford: University Press, 1969; reprint ed., London: Museum of London, n. d.

A Veteran of Smokedom. *The Smoker's Guide, Philosopher and Friend*. London: Hardwicke & Bogue, 1876.

Bastien, André Paul. *La Pipe*. Paris: Éditions Payot, 1973.

―――. *Von der Schoenheit der Pfeife*. Translated by Dr. Brigitte Fischer-Hollweg. Muenchen: Wilhelm Heyne Verlag, 1976.

Belloncle, Michel. *Les Objets du Fumeur*. Paris: Librarie Gruend, 1971.

Bevan, Samuel [Cavendish]. *To All Who Smoke! A Few Words in Defence of Tobacco; or, a Plea for the Pipe*. London: Baily Brothers, 1857.

Bibliotheca Nicotiana; A First Catalogue of Books About Tobacco. Birmingham: Josiah Allen, 1874.

Billings, E. R. *Tobacco: Its History, Varieties, Culture, Manufacture and Commerce*. Hartford: American Publishing Company, 1875; reprint ed., Wilmington, DE.: Scholarly Resources Inc., 1973.

Brand, Alison, comp. *Antiques*. New York: Galahad Books, 1974.

Brennan, W. A. *Tobacco Leaves*. Menasha, WI.: George Banta Publishing Company, 1915.

Brongers, Georg. *Nicotiana Tabacum*. Amsterdam: H. J. W. Becht Uitgeversmaatschappij, 1964.

―――. *Van Gouwenaar tot Bruyère Pijp*. Amerongen: Uitgeverij W. Gaade B.v., 1978.

Buzzati, Dino e. Ramazzotti, Eppe. *Il Libro delle Pipe*. Milano: Ed. Antonioli, 1946; reprint ed., Milano: Ed. Aldo Martello, 1966.

Cardon, E. *Le Musée du Fumeur*. Paris: Maison E. Cardon et Illat, 1866.

Chute, Anthony. *Tabaco*. London: Adam Flip, 1595; reprint edited by F. P. Wilson Oxford: Basil Blackwell & Mott, Ltd., 1961.

Cole, J. W. *The GBD-St. Claude Story.* London: Cadogan Investments Ltd., 1976.

Cope's Tobacco Plant. *Pipes and Meerschaum, Part the Second. The Pipes of Asia and Africa.* Cope's Smoke Room Booklets Number Eleven. Liverpool: 1893.

Cundall, J. W. *Pipes and Tobacco Being a Discourse on Smoking.* London: Greening & Co., Ltd., 1901.

d'Allemagne, Henry René. *Decorative Antique Ironwork. A Pictorial Treasury.* New York: Dover Publications, Inc., 1968.

Davey, Peter, ed. *The Archaeology of the Clay Pipe, I. Britain: The Midlands and Eastern England.* British Archaeological Reports 63. Oxford: British Archaeological Reports, 1979.

Dunhill, Alfred. *The Pipe Book.* London: A.&C. Black, Ltd., 1924; revised ed., London: Arthur Barker, 1969.

Ehwa, Carl Jr. *The Book of Pipes and Tobacco.* New York: Ridge Press, Random House, 1974.

Encyclopédie du Tabac et des Fumeurs. Paris: Éditions du Temps, 1975.

Ewers, John C. ed. *Indian Art in Pipestone. George Catlin's Portfolio in the British Museum.* Washington: Smithsonian Institution Press, 1979.

Fairholt, F. W. *Tobacco: Its History and Associations.* London: Chapman and Hall, 1859; reprint ed., Detroit, MI.: Singing Tree Press, 1968.

Flood, R. J. *Clay Tobacco Pipes in Cambridgeshire.* Cambridge: The Oleander Press, 1976.

Forrester, Henry [Alfred Crowquill]. *A Few Words about Pipes, Smoking & Tobacco.* Publication No. 1, Arents Tobacco Collection. New York: New York Public Library, 1947.

Fraikin, Jean. *La Fabrication de la Pipe en Terre.* Liège: Éditions du Musée de la Vie Wallonne, 1978.

Fume, Joseph. *A Paper:—Of Tobacco; Treating of the Rise, Progress, Pleasures, and Advantages of Smoking.* London: Chapman and Hall, 1839.

Haeberle, Adolf. *Die Beruehmten Ulmer Maserpfeifenkoepfe.* Amberg/Oberpfalz: Verlag Otto Wirth, 1950.

Hamilton, A. E. *This Smoking World.* New York: The Century Company, 1927.

Hamilton, Henry W. *Tobacco Pipes of the Missouri Indians.* Memoir Number 5. Columbia: Missouri Archaeological Society, 1967.

Harley, Laurence S. *The Clay Tobacco-Pipe in Britain.* Essex: The Essex Field Club, 1963; reprinted ed., with addition, Essex: The Essex Field Club, 1976.

Helme, D. *The Clay Tobacco Pipe. An Illustrated Guide.* Durham: Brian J. Hewitson, 1978.

Heward, Edward Vincent. *St. Nicotine of the Peace Pipe.* London: George Routledge & Sons, Ltd., 1909.

Hinds, J. I. D. *The Use of Tobacco.* Nashville: Cumberland Presbyterian Publishing House, 1882.

Hochrain, Helmut. *Das Grosse Buch des Pfeifenrauchers*. Muenchen: Wilhelm Heyne Verlag, 1973.

Honey, W. B. *Dresden China*. New York: Tudor Publishing Company, 1946.

Hume, Ivor Noel. *A Guide to Artifacts of Colonial America*. New York: Alfred A. Knopf, 1970.

Jackson, R. G., and Price, R. H. *Bristol Clay Pipes. A Study of Makers and their Marks*. Research Monograph No. 1. Bristol: City Museum and Art Gallery, 1974.

Jean-Léo. *Les Pipes en Terre Françaises du 17me Siècle à Nos Jours*. Bruxelles: Le Grenier du Collectionneur, 1971.

King, J. C. H. *Smoking Pipes of the North American Indian*. London: British Museum Publications Ltd., 1977.

La Pipe Bruyère. Saint-Claude 1856–1956. Saint-Claude: Imprimerie Moderne du Courrier, 1956.

Larsen, Ole W. Ø. *Gamle Piber*. København: Forlaget Sesam A/S, 1978.

Laufer, Berthold. *Tobacco and Its Use in Asia*. Anthropology Leaflet 18. Chicago: Field Museum of Natural History, 1924.

Linton, Ralph. *Use of Tobacco Among North American Indians*. Anthropology Leaflet 15. Chicago: Field Museum of Natural History, 1924.

McGuire, J. D. "Pipes and Smoking Customs of the American Aborigines, based on Material in the U.S. National Museum," *Annual Report of the U.S. National Museum, 1896–97*. Washington: Government Printing Office, 1899.

Mackenzie, Compton. *Sublime Tobacco*. New York: Macmillan Company, 1958.

Mann, Jenny E. *Clay Tobacco Pipes from Excavations in Lincoln 1970–74*. Monograph Series Vol. XV–1. Lincoln: Lincoln Archaeological Trust, 1977.

Marfurt, Luitfrid et Susini, Jean. *Pipes du Cameroun*. Cahors: Imprimerie Tardy Quercy, 1967.

Nihon Sembai Kōsha. *Nihon no Kitsuen-Gu*. Tokyo: Sembai Jigyō Kyōkai, 1966.

Oswald, Adrian. *Clay Pipes for the Archaeologist*. British Archaeological Reports 14. Oxford: British Archaeological Reports, 1975.

———. *English Clay Tobacco Pipes*. London: British Archaeological Association, 1967; reprinted ed., London: Museum of London, n.d.

Penn, W. A. *The Soverane Herbe*. New York: E. P. Dutton & Co., 1901.

Pritchett, R. T. *Historical and Ethnographical Smokiana*. London: Bernard Quaritch, 1890.

Ramazzotti, Eppe. *Introduzione alla Pipa*. Milano: Ed. Aldo Martello, 1967.

Rapaport, Benjamin. *A Tobacco Source Book*. Long Branch, N.J.: By the Author, P.O. Box 84, 1972.

Raufer, G. M. *Die Meerschaum und Bernsteinwaren Fabrikation*. Wien: A. Hartleben Verlag, 1876.

Rhein, J.-Ch. *L'Art de la Pipe*. Genève: Cabinet d'Amateur, 1978.

Rutsch, Edward S. *Smoking Technology of the Aborigines of the Iroquois Area of New York State*. Cranbury: Associated University Presses, Inc., 1973.

Sant'Ambrogio, Diego. *Pfeifen*. Zuerich: Werner Classen Verlag, 1967.

———. *Pipe*. Milano: Ed. Rizzoli, 1966.

Sassi, Anthony L. Jr. *Notes on the Amerind Manufacture of Smoking Devices as Artistic Expression in Northeastern Iroquoia*. Sarasota Springs: Stillwater Press, 1968.

Smokin' Clean. New York: Japan Gallery, n.d.

The Great Exhibition. London 1851. New York: Bounty Books, 1970.

Thomas, T. A. *Praktische Anleitung Meerschaumene Pfeifenkoepfe zu Verfertigen*. Erlangen: Johann Jakob Palm, 1799.

Tobacco Whiffs for the Smoking Carriage. London: Mann Nephews, 1874.

Vincent-Courtier, L. *Saint-Claude et l'Industrie de la Pipe*. Saint-Claude: Imprimerie Moderne, 1921.

Vogel, Carl Adolf. *Jagdschaetze im Schloss Fuschl*. Muenchen: Droemer Knaur, 1974.

von Schwind, Moritz. *Album fuer Raucher und Trinker*. Stuttgart: 1906.

———. *Almanach von Radierungen*. Zuerich: Verlag J. Veith, 1844.

———. *Rauchgebilde-Rebenblaetter*. Zuerich: Rotapfel Verlag, 1952.

Walker, Iain C. *Clay Tobacco-Pipes, with Particular Reference to the Bristol Industry*. 4 vols. Ottowa: Parks Canada, 1977.

———. *The Bristol Clay Tobacco-Pipe Industry*. Bristol: City Museum and Art Gallery, 1971.

West, George A. *Tobacco, Pipes and Smoking Customs of the American Indians*. Vol. XVII. Milwaukee: Public Museum of the City of Milwaukee, June 11, 1934; reprint ed., 2 vols., Westport, CN.: Greenwood Press, 1970.

PERIODICALS

"A Half Million in Pipes." *Pipe Lovers* (June 1947): 176–77.

Allen, Frederick. "Better than Blue Chips: Objects to Appreciate." *New York Magazine* (February 26, 1979):80.

Brice, Carl E. "Pipes at Auction." *Pipe Lovers* (June 1947):170–71, 188.

"Buys Pipes He Won't Smoke." *Pipe Lovers* (December 1948):370.

de Chazal, Christopher. "Calling Poetic Pipemen." *Pipe Line* (Autumn 1972):10.

Duncan, E. Reid. "Mysteries of Meerschaum." *Pipe Lovers* (January 1949):7–9.

Edings, C. A. "The Thornton Wills Collection of Tobacco Pipes." *Connoisseur* (April 1931):230–34, 237.

Exner, Julian. "The Dunhill Shrine." *Pipe World*, British Edition (May 1970):29–32.

Fresco-Corbu, Roger. "Faces on French Clay Pipes." *Country Life* (June 14, 1962: 1445–46.

———. "German Porcelain Pipes." *Collectors Guide* (July 1972):60–63.

———. "How To Smoke An Economical Cigar." *Country Life* (April 1, 1965):747–48.

———. "Pipes for the Military Historian." *Country Life* (June 8, 1967):1474, 1477.

———. "The Art of the German Porcelain Pipe." *Country Life* (February 28, 1963): 424–25.

———. "The Era of the Meerschaum Pipe." *Country Life* (November 10, 1960): 1100–1.

———. "The Many Faces of the Wooden Pipe." *Country Life* (October 3, 1948): 846, 849–50.

———. "The Rise and Fall of the Clay Pipe." *Country Life* (May 21, 1964):1286, 1289.

Garrett, Wendell D. "Paraphernalia of Smokers and Snuffers." *Antiques* (January 1968):104–8.

Gebauer, Paul. "Cameroon Tobacco Pipes." *African Arts* (Winter 1972):28–35.

Giaccio, John. "First Take an Old Hacksaw. . ." *Wonderful World of Pipes* (Vol. 1, No. 2, 1971):16–21.

Hale, H. C. Jr. "The Heide Collection." *Pipe Lovers* (June 1948): 176–78.

Harte, J. "The Famous Fischers." *Pipe Lovers* (March 1948):78–79, 94.

Hughes, G. Bernard. "The Clay Tobacco–Pipe Maker." *Country Life* (December 28, 1961):1633–34.

Kennedy, Charles Edward. "Chinese Pipes for Tobacco and Opium." *Antiques* (March 1969):408–9.

"Larsen Collection Records Danish Tobacco Tradition." *Smokeshop* (October 1977):42.

Machray, Robert. "The Pipes of All Peoples." *Cassell's Magazine* (1902):210–15.

"Man and His Pipe." *Imperial Tobacco Group Review* (May 1971):18–21.

"More than a Pipe Dream. The U.S. Tobacco Museum." *U.S. Tobacco Review*, 20–23.

Morris, Fritz. "The Making of Meerschaums." *Technical World* (April 1908):191–96.

Nelms, S. R. "A Dealer Tells About Young Smokers." *Pipe Lovers* (October 1948): 299, 318.

Oswald, Adrian. "Tobacco Pipes." *Connoisseur Concise Encyclopedia of Antiques*, IV (1959):201–7.

Ramazzotti, Giuseppe. "Classical Clays." *Pipe World*, USA Edition (Autumn 1969):24–25, 27.

———. "In a Harem of Beauties." *Pipe World*, British Edition (May 1970):21–23.

———. "Light a Meerschaum . . . Light as Sea Foam." *Pipe World*, British Edition (May 1970):24–27.

———. "Painted Porcelain Pipes . . . Strictly Collectors' Items." *Pipe World*, North American Edition (March 1971):9–11.

Rapaport, Benjamin. "A Brief Guide to Tobacco Museums in Western Europe." *The Antique Trader Weekly* (October 15, 1974):44–46.

———. "Another Look at American Tobacco Collections and Exhibits." *The Antique Trader Weekly* (January 11, 1978):48–49.

———. "Collector Gives Basics for Tobacco Collecting." *The Antique Trader Weekly* (December 5, 1972):115.

———. "Pipe Collecting Has Worldwide Appeal." *Collectors News* (September 1976):1, 12–13.

———. "Polychrome Porcelains and Pottery Puzzles." *The Antique Trader Weekly* (September 24, 1974): 60–61.

———. "Puffing Potpourri: An Historical and Pictorial Panorama of Pipe Collecting." *The Antique Trader Weekly* (June 29, 1976): 50–53.

———. "The Basics of Tobacco Collecting." *National Antiques Review* (March 1973):15–17.

———. "The Clay Pipe: Collecting Cutties and Churchwardens." *The Antique Trader Weekly* (August 13, 1974):54–55.

———. "The Heavenly Heath Tree—Briar Pipes." *The Antique Trader Weekly* (October 22, 1974):48–49.

———. "The Magnificent Memorable Meerschaum." *The Antique Trader Weekly* (June 18, 1974):58–59.

———. "The Smoker—Collector." *Hobbies* (April 1973):155.

———. "Tobacciana—More than Pipe Smoking." *Collectors News* (July 1973):1, 4, 21.

———. "Tobaccollections: A Guided Tour of Pipe, Pouch and Printed Matter in North America." *The Antique Trader Weekly* (August 19, 1975):42–44.

———. "World is Full of Clubs Collecting Tobaccoana." *Collector's Weekly* (November 21, 1972):1–2, 6.

"Rich, Rare and Unique." *Pipe Line* (Spring 1972): 6; (Summer 1972):8.

Sampson, O. W. "The Geography of Pipe Smoking." *The Geographical Magazine* (August 1960):217–30.

Savinelli, Achille. "A Smoker's Idea of Paradise." *Pipe World*, USA Edition (Autumn 1969):34–36.

Stephan, Fritz. "The Pipe that Started the Maximilian Tragedy." *Hobbies* (April 1934):15.

"The Heide Pipe Collection." *Hobbies* (October 1946):19.

"The History and Mystery of Tobacco." *Harper's New Monthly Magazine* (June 1855):1–18.

"The World of Pipes." *UST Eagle* (Winter 1975):3–7.

Thiede, Hermann. "Reservisten-Pfeifen immer Beliebter Nostalgie—Nostalgie." *Pipe Club* (Nr. 4/1977):4, 6.

INDEX

A

Abguss, 39
African Pipes (See Chapter 7)
Amber, 67, 74, 80, 106, 124–125, 165, 179, 183, 228, 234
Ambroid, 74, 80
Ambroide, 74
Ambrosine, 74
American Indian Pipes (See Chapter 7)
Andrassy, Count, 49–50
Antique Pipe Shops
  Ampersand, 7
  Astley's, 211
  Brian Tipping, 7
  Collector Clay Pipes Co., Ltd., 7
  Denise Corbier, 7
  Mantiques, 7
  The Old Order, 7
  The Pipes Man, 7
Argillite, 156–157
Aufsatzpfeife, 93, 95
Avignon, 22

B

Barton, J. Trevor, 201, 238
Beck, Aaron, 228
Biedermeier, Gottlieb, 39, 42, 130
Birmingham, 191, 197, 211
Bonaparte, Napoleon, 104–105
Books, Reference, (See Chapter 12)
Bragge, William, 191–193, 196–197, 210, 223, 237–238
Brampton Ware, 33–34, 192
Bristol, 178–179, 181, 211, 241–242

C

Calumet, 151, 153, 155, 159–160, 195–196
Cameroon, 165, 168–169, 241
C.A.O. Products, 229–235
Catlinite, 152–153, 159–160, 162
Champlévé, 140, 144, 148, 197

Chinoiserie, 39–40, 42
Cigarette Cards, 10–14
Clark, Mike, 58, 64, 67, 226
Clay (See Chapter 2), 194
Clay Pipe-Makers
  Baernelts (Barentz), William, 17–18
  Belle, Victor, 23
  Blake, William Thomas, 23
  Bonnaud, Hippolyte Léon, 23, 28
  Chockier, 31
  Christie, William, 218
  Cropp, Charles and Sons, 23
  Duméril, 22, 219, 242
  Fiolet, Louis, 22, 219
  Gambier, 20, 22, 30, 242
  Job Clerc, 24–27, 242
  Strong, Sampson, 218
Clay Pipe Styles
  Alderman, 15
  Churchwarden, 15, 18, 173, 174
  Coffeehouse, 23, 29
  Cutty, 23
  Debrecen, 23, 29
  Elfin, 15
  Fairy, 15
  Figurals, 19–20, 22, 30
  Folianten, 18
  London Straw, 15
  Schemnitz, 29
  Yard, Yard of Clay, 15
Cleveland Public Library, 237
Clift, Ed, 99, 226
Cloisonné, 141, 144–146, 148, 197
Collections, by Title
  Arents, George Jr., 237
  Borkom-Riff Collection, 210
    International Collection, 34, 207, 210
  Christy, 210
  Dick, James Lee, 209
  Duckhardt, 207
  Ehrlich, David P., 62, 201, 207
  Gries, Robert Hays, 237
  Half-and-Half, 207
  House of Pipes, 211
  Leavitt and Pierce, 201, 207
  Marxman Heirloom, 112, 116, 118, 201

(Collections by Title, *continued*)
  Metromedia, 73, 87, 124, 184, 208
  Oldman, 156, 158, 210
  Pijpenkamer Icon (Icon Clay Pipe Museum), 221
  Pinto, Edward H. Collection, 211
  Rhien, Jean-Charles, 130, 135, 143, 223, 239
  Ries, Iwan Company, 58, 208-209
  Rothschild, 33, 100, 104, 107, 197, 219
  Squier, E. G.-Davis, E. H., 210
  Thursfield Collection of Broseley Clay Pipes, 218
  Tinder Box International Ltd., 82, 85, 94, 143, 168, 182, 184, 209
  West, George A., 209, 240
  Willett, 34
  Wills, W. D. and H. O. Collection of Tobacco Antiquities, 34, 65-66, 128, 158, 160, 166-167, 179, 211
Confrérie des Maîtres—Pipiers, 219
Crystal Palace Exhibition, 22, 53, 69
Cullman, Howard S. Library, 237

D

Debrecen, 23, 29, 50, 96-97, 99, 173
Deckel, 43
Defregger, Franz von, 42
Düco, Don, 221, 242
Duke of Sussex, 116, 120, 191
Duncan, E. Reid, 50-51
Dunhill, Alfred, 58, 238

E

Erickson, Ken, 31, 64, 79, 136, 167, 226
Ethnographica
  Africa (See Chapter 7)
  American (See Chapter 7)
  Far East (See Chapter 6)
  Near East (See Chapter 6)

F

Fairholt, F. W., 30, 124, 182
Far East Pipes (See Chapter 6), 194
Farouk I, King, 198-199
Fischer,
  Arthur C., 63-65
  August, 63-64
  Gustave, 63-64

(Fischer, *continued*)
  Gustave, Jr., 61, 200
  Gustave, Sr., 61-63, 199-201
  Otto, 63, 65
  Paul, 65
Forrester, Henry, 123
Fresco-Corbu, Roger, 42, 65
Friedman, Martin, 55, 70-71, 88, 137, 171, 178, 226
Fry, William P. "Bill", 241
Fume, Joseph, 133

G

Gesteckpfeife, 39, 93, 95, 97, 175
Givet, 22
Glass Pipes, 178-181, 194
Golden Horn Meerschaum Pipes, 228
Gouda, 18, 220
Gourd, 166, 186

H

Heide, John F. H., 34, 61, 70, 114, 137, 139, 147, 178, 197-198
Hobbies Magazine, 61, 139, 198
Horne, Bob, 228
Hornepipe, 228

I

Imhoff, Wilhelm Pipe Company, vi, 46-48
Imitation Meerschaum, 8, 69, 71, 73, 80, 86
Imperial Tobacco Group Ltd., 188-189
Imperial Tobacco Products Limited, 210
Indian Pipe Classification, 152 ff.
Indian Tribes
  Blackfoot, 153
  Cherokee, 155, 208
  Chippewa, 152
  Eskimo, 156, 158
  Haida, 157
  Iowa, 153
  Iroquois, 156, 241
  Micmac, 153
  Oneota, 153
  Osage, 153
  Oto, 153
  Plains Cree, 153
  Plains Ojibwa, 153-154, 160
  Sioux, 159-160, 162

(Indian Tribes, *continued*)
  Wyandot, 154
Irving, Anthony, 211
Ivory, 106, 144, 147-149, 157-158, 163, 169, 178, 186, 192, 194-195

J

Jackson, Andrew, 191
Jadeite, 144
James I, King 17, 197
Japanese Tobacco Monopoly Corporation, 240
Jasperware, 36-38, 40
Jean-Léo, 22-28, 30, 242
Johnson, Dick, 228
  Johnson's Pipe Shop, 228

K

Kaffir, 165, 177, 195
Kalmasch, 51, 96-97, 103
Keen's English Chop House, 225
Kopf, 39
Kowates, Karol, 50

L

Landerman, Irving, 70, 90-91, 103, 186, 226
Larsen, Ole, W. Ø., 224, 240
Le Brigand au Repos, 50-51
Leonard, Maurice, 72
Linger, B. Company, 79-82
Lowe, J. H., 227-228

M

Maîtres-Pipiers, 219
Marx, Robert L., 112, 118
Maserkopf, 93
Maximilian, Emperor, 58, 196
Maximilian, King, 139
Meerschaum (See Chapter 4)
Meerschaum Pipe Carvers
  Adler, 57
  Au Pacha, 57
  Barling, Benjamin and Sons, 53, 211
  Bartoleme, 57
  Bolzau, Louis, 53
  Cardon and Company, 53, 69
  Czapek, Emanuel, 57

(Meerschaum Pipe Carvers, *continued*)
  Czapek, M., 57
  Demuth, William and Company, 57, 63, 199, 207
  Dolezal, Josef, 57
  Fischer (See Fischer)
  Flegel, Edoardo, 57
  Gambarini, L., 57
  GBD, 53, 74, 86
  Goltsche, 56
  Guyot, H. G., 56
  Hartmann, Ludwig, 57
  Heimann, M., 56
  Held, M., 53
  Held, M., 53
  Hellman, Jakob, 51
  Hiess, Carl, 57
  Hiess, Franz, 57
  Isberg, Helena Sofia, 57
  Kaldenberg, F. J., 57, 71, 82-85
  Kopp, Georg, 57, 134
  Kutschera, Carl, 60
  Lux Brothers, 53
  Macropolo, D., 57, 70
  Mathisse, 53
  Medetz, 57, 87
  Nolze, 56
  Perkins, H., 57
  Rabe, E. R., 57
  Reischenfeld, 56
  Schilling, Heinrich, 57
  Schnally, 57
  Schneider, Arthur, 57, 67
  Simeron, 57
  Skopec, 56
  Sommer, J. Brothers, 56, 58
  Strassner, Michael, 51
  Wegner, Johann Wolfgang, 51
  Ziegler, M., 56
Meissen, 37, 40, 42, 130
Metal Pipes (See Chapter 8), 194
Minet and Roussel, 22
Mundstueck, 39
Murphy, Doug, 143, 209, 228-235
Museums, Public
  Abbey House Museum, 218
  American Museum of Natural History, 207
  Belfast Museum, 173
  Berne Historical Museum, 197, 223
  Birmingham City Museum and Art Gallery, 211
  Blackmore Museum, 210
  Boston Museum, 218

(Museums, Public, *continued*)
British Museum, 152, 154–158, 160–161, 210, 241
Brooklyn Museum of Art, 207
Castle Arch, 218
Castle Museum, Colchester and Essex, 218
Castle Museum, Norwich, 218
Castle Museum, York, 218
Cayuga Museum of History and Art, 207
Cecil Higgins Art Gallery, 218
Château Dufresne, 209
Clive House Museum, 218
County and Folk Museum, Cambridge, 218
County Museum, Taunton Castle, 218
d'Eger, 224
Duke Homestead, 208
Duke University Art Museum, 73, 87, 124, 184, 207–208
Farm Museum of Landis Valley, 207
Focke Museum, 219
Fondation Alfred and Eugénie Bauer-Duret, 223
Germanisches National Museum, 219
Gorey Castle Museum, 218
Grantham Museum, 218
Heimat Museum, 219
High Street Museum and Christchurch Mansion, 218
Horniman Museum and Library, 218
Huntly House, 218
Indian Shrine Association, 209
Jaegerspris Palace, 224
King Museum, 207
Kingston-Upon Thames Museum, 218
Kinosha County Historical Society Museum, 209
Lightner Museum, 208
Lincoln City and County Museum, 218
McCord Museum, 209
Metropolitan Museum of Art, 207
Milwaukee Public Museum, 209, 240
Musée de Ferronnerie, Le Secq des Tournelles, 218
Musée de Grasse, 33, 100, 104, 107, 197, 219
Musée de la Vie Wallonne, 223, 242
Musée de Saint-Omer, 219
Musée des Arts et Traditions Populaires, 218
Musée du SEITA, 20–21, 30, 38, 44, 57, 75, 106, 109, 113, 129, 132–133, 147, 176, 187, 218

(Museums, Public, *continued*)
Museo del Pueblo Español, 219
Museo Federico Marés, 219
Museum and Art Gallery, King's Lynn, 218
Museum and Art Gallery, Petersborough, 218
Museum of London (London Museum), 16, 210
Museum of Mankind, 210
Museum of the American Indian, 207
Museum of the Cherokee Indian, 208
National Museum, Helsingfors, 224
National Museum of Budapest, 50, 224
National Museum of Czechoslovakia, 244
National Museum of Iceland, 224
National Museum of Man, 210
National Museum of Tokyo, 224
Norris Museum, 218
North Carolina Museum of History, 208
Old Salem Inc., 208
Pioneer's Museum, 209
Pitt Rivers Museum, 218
Provincial Museum of Alberta, 210
Salisbury and South Wiltshire, 218
Smithsonian Institution, 207, 240–241
Stanford University Museum of Art, 148
Stewartry Museum, 218
St. Petersburg Historical Society Museum, 208
University Museum of Archaeology and Ethnology, 218
University of Oxford, 218
Valentine Museum, 199, 207
Wilberforce House, 218
Wisbech and Fenland Museum, 218
Yesterheim Norwegian-American Museum, 209
Museums, Tobacco
Austria Tabakwerke, 66, 77–78, 102, 120–122, 125–127, 223
Barling, 211
Brinkmann, Martin, AG, 219
Carreras Rothmans Ltd, 211
DeMoriaan (Blackamoor), 220
Deutsches Tabak und Zigarrenmuseum, 219
Douwe Egbert's, 18, 68, 92, 95, 109, 130, 220
Dunhill, Alfred, 135, 164, 211–217
Gubbhyllan, 224
Haus Neuerburg, 219
Maison Heintz van Landewyck, 221
Musée de la Pipe, 98, 110–111, 117, 219
Musée d' Intérêt National du Tabac, 31, 112, 219

(Museums, Tobacco, *continued*)
Musée du Tabac et du Folklore, 223
Museum Chacom, 114–116
Niemeyer Nederland's Tabacologisch Museum, 220, 239
Oy Strengberg AB Tobacco Corporation, 224
Reemtsma, H. F. and Ph. F., 40, 75, 97, 186, 220
Rossem, van J. & A. C. Company, 220
Stokkebye, Erik, A/S Cigar and Tobaksfabrik, 224
Storm, P., 224
Tobacco-Textile Museum, 207–208
U.S. Tobacco Company, 19, 32–33, 36, 41, 56, 61-63, 76, 90, 103, 112, 116, 118, 155, 158, 162–163, 166, 174, 180–181, 201–207
White Ox, (De Witte Os), 220

N

Naumoff, Charles P., 35, 38, 45, 101, 105, 108, 114, 147, 171, 175, 177, 185, 226
Near East Pipes
Chibouque (Chibook, Tchibouk, Tchibukdi, Tschibouk) 29, 73, 123–125, 175, 208
Copoq, 125, 177
Hookah (Huqqah), 123, 126, 132–133, 173, 194
Hubble-bubble (Gurgarri, Gurguru), 123, 133, 136, 165, 194
Kallian (Kalian, Qalyan), 71, 126, 128, 132–133
Nargileh (Nargeeleh, Narghile) 71, 125–127, 132, 206
Shisheh (Sheeshé, Sheesheh), 132
New York Public Library, 237
Nose Warmers, 182

O

Opium and Opium Smoking, 126, 137, 144, 147–149, 177, 194–195
Opperman, J. E., 179
Oriental Pipes (See Chapter 6), 194
Oswald, Adrian, 17, 210, 241

P

Paktong, 144, 144–145, 147, 177

Parianware, 44–45
Parke-Bernet 84, 34, 61, 67, 71, 88, 119, 145, 148, 171, 198–200, 240
Pipe Collections (See Chapters 9 and 10)
Pipe Collectors (See Chapters 9 and 10)
Pipe Dream Restaurant, 225
Pipe's Tavern, 225
Pipe Substances, North American, 152 ff.
Porcelain (See Chapter 3), 194
Porcelain Manufacture, 37–42
Porcelain Pipe-makers
Bustelli, Franz Anton, 39
Ehder, Johann Gottlieb, 37
Porcelain Pipe Styles, 42 ff.
Pozito, José, 94, 168, 182, 184, 219
Prattware, 32, 34, 178
Puget, Louis Pierre, 50–51

R

Racinet, M. A., 10–11
Ragoczy, 50, 96–97, 173
Ramazzotti, Giusseppi, 22, 239
Ratner, Herbert G. Jr., 60, 90, 199, 226
Repair (Chapter 11)
Reservistenpfeife (Regimental),
Danish 44–45
German, 43–44, 138, 183, 241
Reservoir, 42
Schwammdose, 39
Suddersack, 39
Restoration (See Chapter 11)
Richelieu, Duc de, 191
Royal Saxon Porcelain Manufacture, 37, 40
Ruhla, 8, 51, 53, 93, 108

S

Schemnitz, 29
Schloss Fuschl, 223
Schwind, Moritz von, 54–55, 101
Scrimshaw, 149, 185–186
Serkallian (Chilloom, Chillum) 127, 129, 132
Shell Pipes, 182, 184
Shepherd Pipe, 69, 73
Sobieski, John, 50
Spontoon, 159, 162
Staffordshire, 33–34, 178
St. Claude, 98, 104, 107–111, 114–117, 219
St. Omer, 22, 219
Svenska Tobaks AB, 224–225

T

Terry, Dave, 140-141, 146, 149, 226
Thiede, Hermann, 73, 100
Thomas, T. A., 53
Tobacco Company Restaurant, 225
Tobacco Merchants Association of the United States, 237
Tomahawk Pipe, 156, 159-162
Tootnague, 144
Trade Pipes, 165, 173-174

U

Ulm, 51, 93, 95-98, 108
U.S. Library of Congress, 240

V

Vámbéry, Professor Arminius 123-124
Van Slaten, 22
Vicarius, Johannes Franz, 22, 39
Vienna, 50, 53, 56, 61, 69, 74, 83-84, 99

W

Water Pipes (see Chapter 6)
Watteville, Baron de, 144, 197, 223
Wedgewood, 36-38
Whieldonware, 33-34, 178
Wilhelm II, Kaiser, vi, 72
Wise Man Pub, 226
Wood (See Chapter 5)
Wood Pipe Carvers
  Bickeling, Johann, 93
  Cutler, Leon, 116
  Davidson, Jo, 116, 118
  Demont, Johann Peter, 93
  Demuth, William & Company, 110, 207
  Drake, Edwin F., 116
  Erdman, Anton, 93
  Frank, S. M. & Company, 110
  Garlow, 116, 119
  GBD, 108
  Gloecklen, Jackob, 93
  Griffin, G. A., 116
  Hartsock, Hetzer, 116
  Howard, Cecil, 116
  Kapp, Frederick, 110
  Kopp, Charles, 116
  Koppenhagen, Joseph, 110

(Wood Pipe Carvers, *continued*)
  Leinberger, Georg Paulus, 93
  Oexlein, Xaverius, 93
  Reese Brothers, 110
  Schenk, Simeon (Simon), 93
  Schneider, Joseph, 121
  Shima, Louis Ted, 116
  Watts, R. D., 116

Other Titles from Schiffer Publishing

Chewing Tobacco Tin Tags, 1870-1930 Louis Storino. Most of these beautiful little pieces of art are over 100 years old and come in various sizes, shapes, and colors. With a listing of over 000 tin tags described and priced, 2000 illustrated tags, plus he many other illustrated and related features, this work will delight the collector.
Size: 6" x 9"   2000 tags in color   128 pp.
Price Guide
ISBN: 0-88740-857-5   soft cover $19.95

Tobacco Tins: A Collector's Guide Douglas Congdon-Martin. This book is the first full-color reference on tobacco tins. Over 1000 tobacco tins illustrated in full color, reveal the designer's and the lithographer's art. In addition, it contains advertising and other ephemera.
Size: 8 1/2" x 11"   Price Guide   160 pp.
ISBN: 0-88740-429-4   soft cover $29.95

Antique Cigar Cutters and Lighters Jerry Terranova & Douglas Congdon-Martin. For the pleasure and convenience of cigar smokers a wide variety of smoking accessories were manufactured, including cigar cutters and lighters. The text places the cutters in their historical context and contains helpful information.
Size: 8 1/2" x 11"   613 color photos   176 pp.
Value Guide
ISBN: 0-88740-941-5   hard cover $69.95

Great Cigar Stuff for Collectors Jerry Terranova and Douglas Congdon-Martin. This is a compendium of cigar related "stuff". Here is the breadth of advertising, ashtrays, matchsafes, cigar boxes, dispensers, and holders that have adorned homes and shops for 100 years and more.
Size: 8 1/2" x 11"   Over 500 photos   160 pp.
Price Guide
ISBN: 0-7643-0368-6   soft cover $29.95

Cigar Box Labels Portraits of Life, Mirrors of History Gerard S. Petrone. Those who fancy these stunningly beautiful paper images as a hobby are in the midst of the hottest area of antique tobacco advertising collecting today. Showcased here are some of the finest and most desirable examples produced by the old stone chromolithographic method between 1860 and 1910. This book also explores some of the rich historical past that surrounds cigars—their manufacture, marketing, and, most of all, their mystique. Also featured is a potpourri of contemporary anecdotes, poems, and other bits of literary whimsy designed to amuse, educate, and titillate the imagination.
Size: 8.5" x 11"   530 color photos   176 pp.
Price Guide
ISBN: 0-7643-0409-7   hard cover $39.95

Camel Cigarette Collectibles: The Early Years, 1913-1963 Douglas Congdon-Martin. This book celebrates the first fifty years of Camel advertising and packaging. Color photographs capture the rich images used to promote Camel goods. The images are accompanied by useful captions and an informative text.
Size: 8 1/2" x 11"   450 color photos   192 pp.
Price Guide
ISBN: 0-88740-948-2   soft cover $29.95

ZIPPO: The Great American Lighter David Poore. This book is a must for all collectors and lovers of Americana. It contains a sequential history of Zippo series, cases, inserts, fluid cans, flint packages, and sundries. It is richly illustrated in full color with many of the most highly prized Zippo lighters that people collect. Information was based on original Zippo salesman's catalogs, leaflets, advertising brochures, and the study of thousands of Zippo lighters. Prices are provided with each picture, and a complete price guide is located in the back of the book.
Size: 8 1/2" x 11"   500 color photos   196 pp.
Price Guide
ISBN: 0-7643-0203-5   hard cover $39.95

Pocket Matchsafes, Reflections of Life & Art, 1840-1920 W. Eugene Sanders, Jr. and Christine C. Sanders. Matchsafes are presented here as a microcosm of life and art from 1840 to 1920. Each is described with details of its pertinent artist, patentee, manufacturer, materials, construction, and value, all complementing the brief and conversational general text.
Size: 8 1/2" x 11"   398 color photos   176 pp.
Price Guide   Index
ISBN: 0-7643-0324-4   soft cover $34.95

The International Collectors' Book of Cigarette Packs Dr. Fernando Righini & Marco Papazzoni. From humble beginnings in the late 1800s, the popularity of cigarettes grew into the twentieth century. Today these miniature works of advertising art have attracted a large, international audience of admirers and collectors. This new book presents hundreds of cigarette packs from every corner of the globe, in full color, with vital information about manufacture and price that will be invaluable to the collector. The images themselves represent a fascinating visual history of more than one hundred years of cigarette manufacture.
Size: 8 1/2" x 11"   700 color photos   176 pp.
Price Guide
ISBN: 0-7643-0448-8   soft cover $29.95

Camel Cigarette Collectibles: The Early Years, 1913-1963 Douglas Congdon-Martin. This book celebrates the first fifty years of Camel advertising and packaging. Color photographs capture the rich images used to promote Camel goods. The images are accompanied by useful captions and an informative text.
Size: 8 1/2" x 11"   450 color photos   192 pp.
Price Guide
ISBN: 0-88740-948-2   soft cover $29.95

Camel Cigarette Collectibles: 1964-1995 Douglas Congdon-Martin. In 1988, R.J. Reynolds began one of the most clever and effective campaigns in advertising history. On its 75th anniversary a suave, new character appeared in the Camel Advertising: Joe Camel, a figure which now has international recognition. In addition to traditional posters and signs, these years saw the introduction of hundreds of premiums and merchandise bearing the Camel logos. Joe was retired in 1997 and the collectibles with his image are more desirable than ever. This is the reference guide with color photos and helpful information.
Size: 8 1/2" x 11"   500 color photos   176 pp.
Value Guide
ISBN: 0-7643-0196-9   soft cover $29.95

Harley Davidson Motorcycles, 1930-1941, Revolutionary Motorcycles and Those Who Made Them Herbert Wagner. Relives the golden age of Milwaukee motorcycling of the 1930s. Hundreds of period photos and a massive text trace the development of the H-D Big Twin from the sidevalve VL to the 61 and 74 models, and the legendary Knucklehead. Experience the Milwaukee motorcycle scene from the men and women who lived it.
Size: 8 1/2" x 11"   354 photos   184pp.
ISBN: 0-88740-894-X   soft cover $24.95

Memorable Japanese Motorcycles: 1959-1996 Doug Mitchel. Examples of first, last, and the most unusual Japanese models to hit American shores are covered in detail with over 450 color photos and accompanying text. Almost every cycle shown is 100 percent original or has been painstakingly returned to its original form.
Size: 11" x 8 1/2"   355 color photos   152 pp.
ISBN: 0-7643-0235-3   hard cover $34.95

Motorcycle Collectibles Leila Dunbar. A cross section of the memorabilia and mementos of the past ninety five years of motorcycling is displayed and discussed in this fascinating book, everything from advertising art and photographs to dealer jewelry and motorcycle toys. Collectibles from legendary companies such as Harley Davidson and Indian abound. Color photographs present the wide range of motorcycling materials available to everyone who hears the call of the open road.
Size: 8 1/2" x 11"   613 color photos   216 pp.
Price Guide
ISBN: 0-88740-947-4   soft cover $29.95

More Motorcycle Collectibles Leila Dunbar. Cycling artifacts in this sweeping survey include advertising, books and magazines, promotional items, Harley and Indian paper collectibles, posters, oil cans, and sales catalogs. This text explores the rich history of the sport and hobby of motorcycling.
Size: 8 1/2" x 11"   600 color photos   192 pp.
Price Guide
ISBN: 0-7643-0333-3   soft cover $29.95

Country Store Antiques: From Cradles to Caskets Douglas Congdon-Martin with Bob Biondi. Included are hundreds of country store items, gathered from some of the best collections in America, and presented in beautiful full color. Also has period photos of actual country and general stores.
Size: 8 1/2" x 11"   481 color photos   160 pp.
Price Guide
ISBN: 0-88740-331-X   soft cover $29.95

Country Store Collectibles Douglas Congdon-Martin, with Robert Biondi. Nearly 600 items, including fixtures, products and advertising from collections across America are illustrated, plus historical photographs of the stores themselves and the people who worked in them.
Size: 8 1/2" x 11"   564 color photos   160pp.
Price Guide
ISBN: 0-88740-274-7   soft cover $24.95

Remember Your Rubbers! Collectible Condom Containers G.K. Elliott, George Goehring, & Dennis O'Brien. This book is indeed a "first," the premier book exclusively about rubber containers. You and your friends can share hours of enjoyment getting to know about one of the hottest antique advertising collectibles ever! You'll learn which brands are rare and which are not. You'll know not to spend a fortune on a "Deans Peacocks" (the "Prince Albert of rubber tins") as well as not to pass on a one-of-a-kind like the "Rainbow," both from the same company!
Size: 6" x 9"   275 color photos   144 pp.
Price Guide
ISBN: 0-7643-0414-3   hard cover $29.95

Schiffer books may be ordered from your local bookstore, or they may be ordered directly from the publisher by writing to:
Schiffer Publishing, Ltd.
4880 Lower Valley Rd; Atglen PA 19310
(610) 593-1777; Fax (610) 593-2002
E-mail: schifferbk@aol.com
Please include $3.95 for the first two books and 50¢ for each additional book for shipping and handling. Free shipping for orders over $100.

Write for a free catalog.
Printed in the United States